Norman Marsh's
Flybox

*For the young and not so young who have
discovered the inexplicable
magic wand.*

Norman Marsh's
Flybox

The
Halcyon
Press

Published by

The Halcyon Press.

A division of

Halcyon Publishing Ltd.

C.P.O. Box 360, Auckland, New Zealand.

Printed by

Colorcraft Ltd,
Hong Kong

Typeset by

Typeset Graphics

ISBN 0-908685-28-9

Contents

Acknowledgments

My heartfelt thanks to Jean Marsh whose patience, correcting my repeated spelling mistakes, must at times have surpassed that of Mother Teresa. Whenever I have returned late, tired, or thoroughly drenched from a fishing expedition I have never had less than a welcome smile. To Norman Nelson Marsh of Wanganui for advice on fishing matters. A special thanks to the following: Peter Davies of Optics, Motueka, photography assistance, fellow members of the Southland Fly Fishing Club and the Motueka Fishing Club, especially John Craze. To Len Prentice of Southland, a great fishing buddy, and members of the Southland Fly Fishing Club. To John Goddard (UK) for many shared adventures, Val and Sue Atkinson of San Francisco premium angling photographers. Rick Schmitt and Mark Freeman of Michigan (USA) wonderful hosts. To my fishing mate Vern Williams for sharing a lot of his fishing life with me. His advice has been invaluable.

No author can claim his book is written unassisted. Especially a fishing book. The advice, suggestions, information, and not least encouragement from many friends, fellow writers, and fishing mates, all are expressed between the following pages. For this I am most grateful.

Norman Marsh

Introduction

After a near lifetime of chasing trout I should know all there is to know about trout fishing. But, as any experienced angler will confirm, it would take more than a lifetime to solve all the riddles. It is a constant learning process and I thank my past fishing friends and acquaintances for passing on whatever angling lore I now possess. I still have lots to learn which, you may think, is hardly a licence to write another book on fly fishing but I have also found that, although there is no substitute for personal experience, there is much to be gained from other anglers.

When first approached by my publisher regarding this book I admit to scratching my head. So much has been written about fly fishing. Then I realised if I am still learning so must others. I hope that in these chapters the reader will find some new ideas, following perhaps an adaption of an existing method but if I do repeat, in some way, a recognised technique I hope it will be of value to the less experienced. I apologise to the knowledgeable angler.

I must have been about 12 before I caught a trout on a fly but long before that many a fish had been given a whack on the head after being tickled from some tiny Yorkshire beck. At not much more than seven years of age I was already adept at poaching but, after being apprehended by a kindly gamekeeper and an even more benevolent landowner, I was shown the light. It may have shone but it was very dim and a further episode saw me before what would now be a children's court receiving a fair warning. For the life of me I couldn't see why anyone could worry about a few trout especially when they weighed about three to the pound but it was a lesson brought home to me when my father stopped my weekly allowance for quite some time.

It must also have taught my father a lesson because not too long afterwards and as it was near my birthday I became the proud, surprised owner of a second-hand split cane rod. Even better, my father put me under the wing of Billy Crabtree, a friend and member of a local fishing club. They had the lease of a tiny stream near Haworth, even then famous as the home of the Bronte sisters, and for five shillings I was allowed to fish the little pools for a whole season. In those days it seemed a lifetime.

The tackle was crude by today's standards and my efforts clumsy but a lasting memory is when a treasured 'Hardy's Favourite' floated near some gnarled tree roots and became snagged. To my utter amazement, as it came free and skated across the surface, a trout of about a half pound (no metric then) grabbed it and I was on the lifelong road of fly fishing.

It was in the mid 70s after near 20 years of fly fishing in New Zealand that I recognised the scarcity of information on trout food insects. Up to that time most anglers had relied on popular patterns, artificials that paid only lip service to our indigenous aquatic insects. Finding that practical information was scarce I spent the next seven years working through existing scientific papers, relating them to fly fishing. After I had spent countless hours collecting specimens on the rivers and streams throughout New Zealand, *Trout Stream Insects of New Zealand* was published by Millwood Press (1983). The book was naturally centred on trout flies, both natural and artificial, with little on the mechanics of fly fishing. This was rectified with the publication of *Troutfishing* by the Halcyon Press (1990).

But time passes and new anglers arrive, each with the same thirst for knowledge that we experienced during our own halcyon days. Fishing techniques and tackle are ever changing. Fly tying is becoming more popular as anglers realise the advantages of 'tying your own' and now with new angling methods even more-so. It is not so long ago that, if asked to explain an 'emerger' or 'compara-dun' we would have been hard put for a reply. Leader design has also changed significantly.

Within the pages of this book will be found sketches and photographs of aquatic insects which the trout and the angler are likely to encounter. Some are common, others only occasionally of note, including seasonal insects which may occur in some regions and not in others. Some are peculiar to certain streamside habitats, others usually appear in the evening hours and some types of fly occur in various guises. Streamside observations are very important — such as noting what time of year a particular insect appears, what time of day and the reactions of the trout. I have tried to cover a fair range of trout stream insects but have omitted rare species, the descriptions of which would only serve to confuse. Also my list of artificials is reduced to a practical level. Not all are my creations because fly fishers, far more skilled than I, have come up with some pretty good imitations that have stood the test of time. When trout are feeding it doesn't take much skill to catch them but when they are feeding selectively — ah, that is another matter!

It is very rare for two fly fishers to agree. Each will fall back on personal experiences and, with everchanging fishing conditions, this is not surprising. One school of thought promotes the idea that it is the presentation of the fly that is most important while another says that, regardless of expertise, it is the fly that counts. After over six decades of chasing trout, I feel sure that the successful fly fisher is the one that marries these two ideals. A working knowledge of a trout's general diet plus good casting skills add up to maximum pleasure when a fly has been chosen to match the natural insect. It needs only a basic understanding of entomology, and I emphasise basic, to provide another string to the angler's bow and one, if plucked correctly, make the trout dance to *your* merry tune.

1

Trout Flies

If there is one thing a fly fisher will never be short of it is the choice of artificial trout flies. There are literally thousands of patterns and they come in all shapes, sizes and colours. The great majority can be classed as 'fancy flies', created from the imagination of the fly tyer who decides, hopefully, that the trout will eat it. 'Fancy flies' have been with us for generations. In fact it was nearly 500 years ago that Dame Juliana Berners, an English prioress, came up with a list of 10 artificial trout flies, all based on natural insects.

Incidentally, the question may well be asked: Why, after this ancient prioress gave us some 500-year-old fly tying lessons, has it taken so long for fly fishing to come to the attention of women? Perhaps it was because in those bygone days only the wealthy could read, and with books then handwritten, it was the learned clergy who, in addition to producing holy scriptures, ventured into other literary works. More likely the Dame had little knowledge of her subject but was simply a ghostwriter for a well-heeled, angler of that period. I think she was a typical woman. Far too busy with practical matters to entertain such a silly pastime as chasing trout. Perhaps the royalties were the attraction. One hundred and fifty seven years later Izaak Walton produced another list, and all based on natural insects. Walton

described his fishing in lyrical terms but, apart from describing trout flies, by his own admission he wasn't much of a fly fisherman and spent most of his days fishing with baits and other unmentionables. Not the least snails with slit bellies or cow-turd beetles.

It was Alfred Ronalds, who gave fly fishers their first scientific classification of trout stream insects. In 1983 I obtained a copy of *The Fly-Fisher's Entomology* (1836) in which he describes, in detail, 47 artificials, all based on naturals. His hand-painted illustrations can only inspire admiration for that talented angler of yesteryear. Ronalds in his introduction, makes the point that 'some anglers seldom make their own flies; and unless this accomplishment has been attained, half the pleasure of fly fishing has not been tasted'. He was also one of the first anglers to graphically describe the trout's 'window', that intriguing feature much discussed among fly fishers.

Nevertheless, I suspect that the great majority of fly fishers are satisfied, however grudgingly, with 'bought' flies. The standard of these vary in the extreme from those tied with poor materials, to top examples of the fly tyer's art, the later obviously hard to come by. While machine-made trout flies are impossible to manufacture they are produced by the thousands from many sources, not the least 'backyard' enterprises using poorly paid, nimble-fingered operatives who have never seen a bright clean river, much less a trout. Judging by the increasing number of extra curriculum fly tying classes the ranks of fly tyers are growing rapidly, not I think, purely for economic reasons, but for the ability to ensure 'quality' artificials and, what is an even greater incentive, the freedom to imitate trout stream insects. The benefits of tying your own can only be appreciated on occasions when, despite a brimming fly box, none of its inhabitants can fool the trout. They may be well tied but the missing fly may be one very suggestive of an insect on the water. What better than an example:

Commercial fly tying circa 1930.

Flytying assembly line.
(note supervisor)

A good fishing friend of mine, whose skill with the rod shames me, fishes throughout the season with a few bought patterns. Together we often fished a small Southland river running through the Hokonui hills, a flax-lined stream blessed with a series of numerous pools, each one tumbling into the other over miniature waterfalls. As we made our way through the paddocks Ben pointed across the river. 'There's a deer!' he exclaimed and, sure enough, a large russet brown animal was loping up through the high tussock. We gazed enthralled at this lucky sighting until, as it reached the top a familiar form took shape. 'It's not a deer, it's a hare!' I replied. It had been a remarkable illusion. Against the steep featureless tussock-clad face and without scale of reference the hare seemed many times larger. It was a day for surprises. Nearing the river and climbing the last hummock Ben nearly stepped on a skylark's nest containing three speckled eggs. So well camouflaged are these nests that chance discovery is rare indeed.

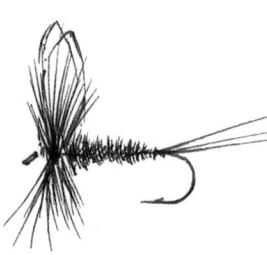

Modern dry fly.

A cloudless sky and a bright sun soon had us shedding jackets and pullovers and after creeping and crawling through bankside scrub and over small bluffs we were thankful for the Lion Brown that Ben had included in his lunch bag. Ben, using a large Coch-y-bondhu, had landed and returned some nice plump trout around the 1kg class, while I had had similar sport using a #12 Kakahi Queen. In the afternoon the trout went berserk. Typical of bare country on a hot day the light breeze strengthened into a half gale and soon the stream was sprinkled with grasshoppers. My fly box soon produced a hopper pattern which duped more trout slurping the insects from the tops of the now well-whipped water.

It became one of those red letter days. Trout in a feeding frenzy slashing as the fly bounced on the surface. When Ben returned from well upstream he had a sorry tale to tell. 'Damned fish were feeding like the clappers on grasshoppers and I couldn't touch them!' Then, eyeing the two large fish lying in the tussock, added, 'You've done all right!' In fact the trout had turned very selective taking only grasshoppers and as he had no imitations he missed out on some fabulous sport. Later that afternoon the wind dropped and the trout returned to sporadic feeding, but not before Ben had taken another fish or two on a borrowed fly. 'I'll keep this as a reminder,' he said as we trudged back to the car and I noticed on future occasions that his fly box contained some very nice grasshopper imitations as a result of attending the local fishing club's tying sessions.

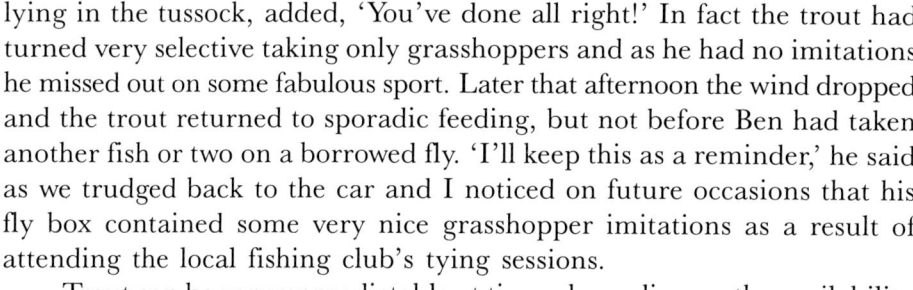

Trout fly old style.

Trout can be very unpredictable at times depending on the availability of food and, even if many popular ready-made patterns look nothing like a natural insect, it doesn't mean they are less effective. A good example is the Royal Coachman. A flashier dry fly you would be hard put to find. It is a great trout catcher but you will also find it much less effective on the less boisterous rivers featuring long, near-still pools, or quiet backwaters. Here the trout has ample opportunity to examine the artificial and compare it with flies commonly taken in its diet.

Some popular patterns, even unintentionally, represent a natural insect. For instance, a Greenwell's Glory roughly imitates some *Deleatidium* genus, especially the olive/brown duns. This English pattern has been around for over 130 years and the fact that it is successful on either side of the globe must indicate that some of our waterside insects are similar to those overseas. A typical English trout fly of a style dating back to Waltonian days is faithfully reproduced even to the whipped gut eye (it was not until the late eighties that practical eye hooks were developed.) These heavily-dressed flies were always fished across and down. Mottram's* imitations, tied during his visit to this country in 1910 shows the rapid advance of style in trout flies. Up to this time the dry fly was little used and indeed scorned by fly fishers of the period. Captain Hamilton, author of *Trout Fishing in Maoriland* (1901) was a prime example and was very forthright in maintaining that

Copy of Early English trout fly.
(note gut eye)

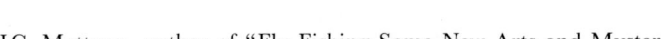

* J.C. Mottram, author of "Fly Fishing Some New Arts and Mysteries" 1915

only five trout flies were necessary to catch trout anywhere in New Zealand. All wet flies.

Creative New Zealand fly tyers have not been idle in imitating natural insects. An early example was the late Basil Humphries, village postmaster. With his creation of the Kakahi Queen, an excellent copy of *C. humeralis*. Others are mentioned in later pages. They have not fallen into the trap of 'exact' imitation where insects are copied slavishly but, in the dressing, emphasise some outstanding feature of the natural whether in shape or colour. A successful fly fisher pictures his or her nymph or dry, not as it appears to him, from above, but from the trouts' viewpoint. Naturally, in this chapter on trout flies the emphasis has been on the fly, but of equal importance is presentation. An angler with good streamcraft and casting skills together with a basic knowledge of entomology (there — I've said the awful word!) will taste pleasures denied mere mortals.

A bag of trout in Captain Hamilton's day. Note the barge pole.

Captain Hamilton's trout flies.

2

Fly Hooks

Reindeer horn fish hook
circa 2000BC.

Wooden hook from Norway.

The perfect artificial fly, one dressed with such skills that it would invariably deceive the most suspicious of trout does not exist, but its partner, the fly hook, can be nothing less. Modern fly hooks are a far cry from hooks of earlier times when Dame Berners (what an enterprising lady!) using needles, provided details of hook making in *Treayste of Fyshinge wyth an Angle* (1496). Now, with the aid of computers and specialised machines, they are mass-produced and generally of excellent quality. Modern hooks are the 20th century's gift to the angler, and the fly fisher in particular.

Gone are the days when hooks were tied to gut, an animal byproduct, a drawn filament that needed soaking before use and deteriorated with age. While primitive examples of eyed iron hooks have long been discovered it was not until the late 19th century that, thanks to one G.S. Marryatt, a contemporary of the now legendary Halford*, that modern eyed fly hooks were developed. Since then the number and style of fly hooks is almost beyond belief. Mustad, a Norwegian company, and one of the world's largest producers of hooks, lists no fewer than 200 varieties ranging from a massive shark hook weighing 3kg to micro fly hooks of which a 1000 weighs less than an air-mail envelope.

The cost of fly hooks compared to the rest of our fly fishing outfit is

16

* F.M. Halford "Floating Flies and How to Dress Them" 1886

infinitesimal. We spend hundreds of dollars in rods, reels, waders, fly boxes, nets, licences, transport and little gadgets eventually left at home, so to economise on hook quality is foolish in the extreme. How often have we reshaped a bent hook or put up with a rusty one. For all our skill and effort to tempt a trout if the hook is faulty or blunt it is a waste of time. We may not be able to catch trout but we certainly don't need to educate them.

REQUIREMENTS OF A FLY HOOK
1. Needle sharpness.
2. A tapered eye, neat and smooth. Either down eye or up, mostly down.
3. A delicate barb for easy penetration.
4. The point straight and with no 'claw bend'.
5. Hook bend is also optional.
6. Shank length should be proportional to the hook size. Twice the gap.
7. Wire size. Standard wire hooks are recommended.

Hook shape is important. Shank bend and gap should be in proportion. Lengthening or shortening the hook shank in relation to the gap will reduce hooking effectiveness. The ideal shape is the round bend which, when buried to the full, gives a good hold. If the hook is slightly offset so much the better. Barbless hooks are becoming increasingly popular, present no greater risk of losing fish, and are much safer to use. A painful visit to the doctor may seem a nuisance, but compared to leaving a stream when trout are on the go, it is a minor disaster! Great care should be taken not to weaken the hook if barbs are crimped down with pliers.

HOOK SIZES
At one time hook sizes used to be very confusing with an Old and a New Scale and now, with the advent of Japanese hooks, it looks like we are back to square one. Mustad use the scale we are most familiar with. This using 1 (one) as a neutral point and using even numbers, hooks growing larger in gap size decrease in number. For instance a #14 hook size is smaller than a #12 and so on. I have never used hook scale for the Japanese product trusting to remember the speci number.

HOOK STYLES
Hook style is really an angler's preference. Of today's styles the forged straight, bronze, turned-down eye is the most popular. Other styles are the Model Perfect, the Round, the Sproat and the Limerick. More important is the correct choice of hook for the purpose. Below is a selection of hooks which, over the years, I have found the most suitable and reliable. To these should be added Kamasan hooks which are chemically sharpened and of fine quality. They are more expensive but, as mentioned earlier, all hooks are relatively cheap compared to the rest of your tackle. For all that I find

Mustad to be of good quality, and if the trout have no preference, when they take off with the fly it is a little less painful.

MY HOOK SELECTION

	Dry fly	Wet fly	Nymphs	Caddis/Grub	Stonefly
Mustad	94840	3906	3666	37160	9672
Dai Riki	305C				
Kamasan				B830	

The Mustad 94840 can be classed as a standard size. Hook sizes are determined by the hook gap and shank length. A Mustad 94840 2XL has the gap of a standard 94840 but the shank is two hook sizes longer than the standard. A 94840 1XL has the gap of a standard 94840 but the shank is one hook size longer than the standard. A Mustad 1XS has the gap of a standard 94840 but is one hook size shorter than the standard. The 94840 standard wire diameter is a good choice for dry flies whereas the heavier wire of the 3906 makes it suitable for wet flies. Similarly the 3666 for nymphs.

Many fly tyers are now using what are termed 'grub' hooks which are ideal for imitations of caddis larva, stoneflies nymphs (which curve naturally when not at rest) and dobsonfly larva. At a recent fishing seminar where I produced some tiny dressed hooks, the smallest being a #28, some of the audience were dubious about their trout holding qualities. I suggested that small is not necessarily weak and made the point (pun unintended) that tempered steel wire, however fine, is tremendously strong and that once the hook bend is fully secured a direct pull on the hook in an effort to stretch the steel, a near impossible task. If the hold is fully secure the only possible way to lose a trout is by the knot or nylon failing or the small hook ripping out.

Returned with thanks.

One of the most memorable trout I ever landed was caught using a #16 hook. It was a sea-run brown in the Waikawa River, a brown stained stream running into the estuary. The average fish for that river would be around three pounds. When that great silvery trout took the small fly in a porpoise roll little did I anticipate the trauma ahead. While playing the fish my thoughts were ever on the 4x nylon and the little hook. The battle which lasted at least 20 minutes is as clear in my mind as if yesterday.

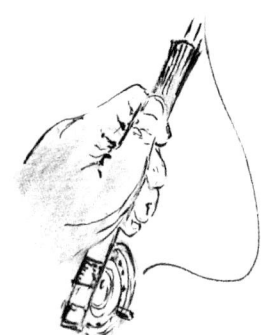

Trout fishermen are like that, even the most absentminded of us can recall long past encounters. At one stage I played that lunging sea-run from the reel spool and had given up the ghost when at the last moment he turned and surged upriver and held in a narrow rock channel. After recovering line and throwing caution to the winds I turned him downstream in the fast current and into the waiting net. My arms and knees were trembling as I weighed him while still in the net. He pulled the pointer down to over 11lb and, despite leaving a liberal sprinkling of silver scales in the net, he shot off back home, a wiser fish I hope. So much for the quality of modern hooks. That hook was still good for the rest of the day's fishing.

The most important feature of a hook is its sharpness. At reasonable intervals the point should be checked, especially after a fish has been landed or, more importantly, if one is lost. There is no need to buy an expensive one. Diamond grit files can be purchased at any pharmacy for little cost, the plastic handle drilled, and some fine cord attached. A suspect hook point should be immediately discarded. Similarly with hooks bent out of shape. Trying to straighten one will certainly weaken it. Overuse of the file is not recommended — one definite stroke along the point is far better than fiddling around. If you have to economise with tackle, never choose the hook.

A blunt hook is unlikely to hook a trout firmly, particularly if the temporary purchase is in the mouth's hard gristle. It can, however, without fail, puncture some part of the angler's anatomy and that hazard is always with us. Indeed, what with the continuous swishing around of the fly at high speed it is surprising we suffer so little. Accidents do occur however and in time, sooner or later, we can expect trouble. My first experience occurred when a nearby angler got a Veltex spinner caught in some weeds and, after giving it an almighty heave, it came hurling back at him at great speed. Instinctively he shielded his face and received the treble hook in the palm of his hand which, to say the least, was inconvenient — the nearest doctor some 50 kilometres away. We discussed riverside surgery but when I drew my somewhat rusty knife from a not overclean pocket he turned pale and headed for his car. Perhaps if I had been familiar with 'press and pull' technique, things may have been different.

Press and pull is a very handy way of removing embedded fish hooks and, if not too popular with the victim, is very effective and almost painless. The problem lies in the barb but by using 'press and pull' it can be retrieved

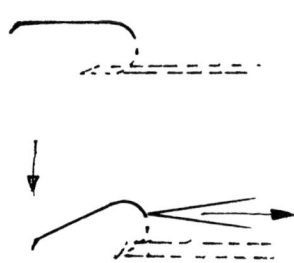

Depressing the eye tilts hook point down. This enlarges the entry wound. The bend shields the barb when withdrawn.

back through the original entry path. Take a short piece of stout nylon or a couple of finer strands around the hook, bend and tension lightly. The victim will soon tell you if it is too much. Press your thumb firmly on the hook eye until the hook shank is cocked up at a 45 degree angle then shout 'there's a good fish!' and pull the nylon smartly. In nearly every case your patient will shower you with praise but if the operation is not quite a success you might consider beating a hasty retreat.

No matter how long you have fished there will always be the occasion when, despite the lessons learnt, some odd situations will occur. In high hopes we set off to fish the Mataura, Len to act as guide and Jean ready with the camera. The river had been above normal the previous day so, by rights, the day should have been in good 'fettle'. (There's my Lancashire upbringing coming out). But our spirits fell as we crossed the Nokomai bridge. Up country, rain and snow melt had obviously occurred and the river was high with that murky green colour we anglers have come to dread. It was little better at Athol and the three fish we had failed to catch the previous day were lying in parts unknown.

It was back down into the gorge where Len suggested a large backwater. Although it was overcast and muggy, it was still pleasant walking across the long paddock and through the old man willows to the leaf-covered pool. Even from a distance we could see a trout circling, little dimples marking his route. He seemed to be in six places at once but an active fish is a good sign and while keeping Trutta from the water's edge I tackled up. A small Pheasant Tail seemed to be a good choice. Kneeling in the soft mud I spent some time trying to outguess him but the water was murky and, without sun, difficult to ambush. On the odd occasion when I dropped it near him he either missed it or ignored the nymph. What to use? This is

where a bit of aquatic insect habitat knowledge came in handy. A small water boatman on the second cast did the trick and, although not all that large, the brown put up a fair fight, finishing with a high leap and freedom. On inspecting the hook I saw the barb gone. 'Poor hook,' I said to Len which caused some amusement. In fact, I wondered if it had been tied on without a point. Points are rarely broken in a fish so maybe the preacher had sinned. Forgive me, father!

One of my strangest experiences relating to hooks was when fishing the Lochy near Queenstown with John Goddard. We had had a good day tramping up from Lake Wakatipu, both rainbows and browns coming to the net. Some lost, some landed. At the last pool before we returned John met his match with a rainbow that refused to be hooked despite repeated attempts. Len and I didn't improve the situation with comments on the Englishman's progress. When the fish finally took the fly, leapt and said farewell, he inspected the hook. We looked on amazed. By some mischance the hook was bent in a circle from hook point to eye, making that rainbow the luckiest fish in the river.

3

Fly Rods and Fly Lines

The evolution of fly rods is a story in itself. The first rods were simply slim branches of willow, hazel, or some other strong but pliable timber but in the sixteenth century the rods were made of lancewood and greenheart. The latter was an exceptionally strong straight grained wood. To cast the fixed lightweight gut or string lines rods of 18 to 20 feet (5.5 to 6 metres) were necessary. By 1800 the search was on for a strong yet light material with rod makers settling on inferior bamboo. This was in vogue for some time but with the discovery of Tonkin cane from Indo China, a much superior bamboo, it became possible to split the hollow tubes into splints and reform them into composite form. By 1855 practical split cane rods were being manufactured in America and an inspection of the cross section of this remarkable material shows why its use became widespread. Although the inner wall of Tonkin bamboo is relatively soft 'pith' it quickly changes to hard springy fibres in the outer shell. These fibres are not only flexible but form a steel hard 'skin'. The central 'pith' is very important, providing sympathetic support to the flexible outer fibres. Together with the development of oiled silk lines (silk allowed ready impregnation), split cane rods produced much greater casting efficiency with the result that five and six metre rods became obsolete. Split cane fly rods had come to stay or

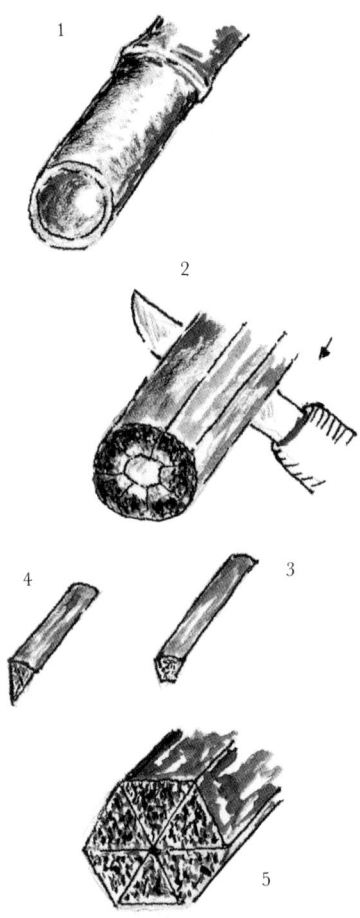

1. Raw Tonkin Bamboo.
2. Cane split into rough sections.
3. Rough splint.
4. Two sides planed. Angles 60°.
5. Glued rod section.

so Victorian fly fishers thought.

One man's vision is another's launch pad. The development of hardened plastic material once again changed angler's ideas of the perfect rod. Fibreglass rods entered the field. The process entailed wrapping woven,

resin impregnated sheets of fibreglass filaments around a tapered stainless steel mandrel, which when baked and the core removed, produced a tubular 'glass' rod. Apart from much improved performance they were relatively inexpensive compared to split cane and a boon to the fly fishing world. Another advantage of the material was its resistance to misuse. While not indestructible (what rod isn't?) they can withstand extremes of temperature and, unlike natural woods, are not subject to warping. Solid glass fibre rods were popular for spin fishing, but it was the lighter tubular rods that became the darlings of the 50s. Fibreglass fly rods greatly increased the popularity of fly fishing and once again anglers thought they had found the perfect material.

I well remember my first fibreglass. A 9' #8/9 Fenwick Ferulite now over 30 years old and hanging on the 'wall of fame' next to some of my other angling memorabilia. Compared to today's rods it feels like a broomstick but it was certainly the rod for handling big Fiordland rainbows. I still use a Hardy's 'Jet', a fibreglass rod, a delight to handle, its softer action ideal for fishing the wet fly.

The search for resilient, stronger, synthetic fibres continued and it was the increased interest in space travel and aeroplane engine research that challenged fibreglass. This new material was carbon fibres, approximately 1/5000th inch in diameter with a single yarn consisting of some 5000 strands. The bonding of these fibres produces a rod with a very high modulus of elasticity in the region of 32 million pounds per square inch (p.s.i.) compared to less than half that for fibreglass. This, plus a high tensile strength and lightness, meant that greater line speeds were available with less 'rod shake' and rod energy could be transmitted to the line more efficiently. The latest (although I am sure there is more to come) in fly rod material is boron, an even stronger (strength for weight) material but which is still combined with graphite in production rods. To the average fly fisher probably the most important aspect of fly rods is the price. Not all of us can afford the top of the range 'graphite' or 'boron' rods but such is the wide range available today that a compromise is easy.

A glance at any fly fishing catalogue is enough to confuse anyone buying a rod. The selection is mind-boggling and, for some of those masterpieces, so are the prices. It is a popular conception that you can buy an allround rod which will be sufficient for all expected needs. While I do not advocate a rack of rods, the one rod syndrome is a fallacy. We have quite a wide range of river types in New Zealand from the broiling waters of the Buller or the Waiau to the gentle spring creeks of the West Coast; from the limpid pools of the Mataura to the boisterous Tongariro and other large North Island rivers. Add to those the many small streams and wilderness waters and it should be clear that the choice of a fly rod is one of the most important decisions a prospective fly fisher can make. He or she should never attempt

Fly rod craftsman circa 1930.

to buy one off the rack without both the advice of a competent fly caster and the opportunity to thoroughly test it. Providing the choice is not outlandish, all fly casters, after a short period, will adapt to a particular rod so there is some tolerance.

A good fly fishing rod should not be a jack of all trades but designed for a specific purpose, either for the dry fly or nymph (stiffish) or wet fly (softer action) and yet when the call comes it should be able to extend its qualities by using one line weight heavier or lighter. For instance on a windless day, with a slight upstream breeze a line weight #6 rod should cast an extended #5 weight line and a 16ft (4.9m) leader with ease. Conversely the same rod should cast a shorter #7 weight line under adverse wind conditions.

The first factor to consider when choosing a fly rod is the type of water to be fished. For small to medium size rivers, where #10 to #18 flies are used, a suitable rod would be a graphite 2.7m (9ft) and rated at #6 for line weight. For fishing on lakes or very large rivers a second rod is necessary. A graphite 2.7m (9ft) #8 is a good choice. Most quality rods will be fitted with a silicone/carbide 'Fuji' butt and end rings but I much prefer stainless steel for the tip, which I find allows for smoother line manipulation. Screw reel fittings are now used on most fly rods — the ones that screw forward from the butt are ideal and worthwhile. If and when the coffers can stand it, a 3m (10ft) #4/5 graphite should be considered. Those long rods are still lightweight and a joy to use on most rivers and ideal for summer dry fly fishing where delicate casting to spooky trout is necessary. A longer rod gives better line control, and it is easier to mend the line for drag free drifts. The action is much slower than for shorter rods but accuracy and presentation is enhanced.

Willow grub time.

My first experience came when I shared one of those long graphites with my mate Vern on the Riwaka River. That river is narrow with more than its share of willow and native bush and, unfamiliar with rod's lazy action I must admit to spending time plucking the fly from trees. It was a sweltering day with a little breeze, and the trout were mostly under or near overhanging trees. Willow grubs, on their tiny threads, were lowering themselves onto the stream surface and the trout were waiting. Most were shy of either the fly or the nylon and up to midday only two trout were landed. In those shrunken pools, if we put one good fish down we must have upset at least a dozen. Around 2pm we rounded a bend and waded across a pool tail. Suddenly we froze. Not 20 paces upstream, lying along a shallow sandy edge with barely enough depth to cover him was a beauty, a brown of at least 2kg.

After an hour, taking turns, and using the smallest of flies, our elation turned to dismay when neither of us could so much as touch him. It was time to change tactics. I cast a #12 nymph well to his right in deeper, faster water. Curious, the fish shifted towards it but then returned to his station. Vern immediately popped a tiny Pheasant Tail upstream and the trout moved up to meet it. Big fish in a tiny river are great fun, especially when you have worked so hard to catch them. Vern's long rod eventually won the battle and a beautifully marked, well-fed fish came to the net. It weighed better than we expected and returned with no real harm done.

During that time of year trout seem to have a fixation on willow grubs and may well need something to shift their attention. Later, as shadows lengthened, caddis flies fluttered crazily across the surface and splashy rises dotted the surface. Under such low water conditions a soft delivery of the leader is vital and by the end of the day I had fallen under the spell of the long rod. St Peter, the patron saint of all anglers, must have been watching. Shortly after that trip my American friend, Mark Freeman, during a visit to 'Streamside', presented me with a 3m #4 Loomis. You see — there is a heaven on earth after all.

Long rods may be great but certainly not for backpacking. This is the domain of the multi-sectioned fly rod, usually with four-piece arm-length sections that mount to make a 2.7m (9ft) rod. Anyone that has toted a long rod case around the New Zealand bush or in an international airport will readily agree. I must admit, though, that most of my long fly fishing life has centred around much less sophisticated rods and looking back I don't think trout noticed the difference.

For wet fly fishing, downstream, with its dreamy action, there is no more delightful rod to use than a #6 old fibreglass. It is very sensitive to the pluck of a sea-run brown that can't quite make up its mind, and together with a razor sharp hook, it sometimes proves their undoing. But for upstream river fishing there is no substitute for a slim powerful graphite with a slightly

'stiff' action, one that can 'punch' #5/6 line into the wind as the occasion demands. Choice of rod or rods depends on the angler's purse and adaptability and if he or she enjoys using a friendly rod, whatever its pedigree and with it catches a fair share of trout, then the finest, most expensive rod in the world, will give no more pleasure.

FLY LINES

Gone are the days when fly fishers relied on a stallion's tail for a fly line. It must have been a painstaking operation using spinning cog wheels to twist and braid fine hair into a line long enough to fish the width of a stream. Stallion tails were chosen. Unlike the mare, a stallion's tail does not come into contact with urine which weakens and stains the hair. Braided silk was also used in bygone days to fashion a fly line and even mixed lines of hair and silk were common. Horse tail hair strands are, more or less, the same thickness and length, and tapered lines were made by twisting say four strands then plaiting these. They were then knotted to sections of similar thickness and the process repeated using only three-strand braided sections; then two, until the tip section became a single strand of horse hair.

In Victorian days the most popular fly line was made of braided silk, impregnated under pressure with oils then coated with special varnishes and dressed much as a French polisher would finish fine furniture. It was a painstaking procedure — with many layers applied and a drying period required between each. The tapering process was a skilled operation reflected in the price which was around £3 in 1923 figures. In recent years there has been renaissance of the silk line, marketed by Phoenix Lines of Sutton Coldfield, England. From braiding to boxing the process takes eight weeks.

In these days of superb plastic lines one might well ask 'Why bother?' For those who have used silk fly lines the answer comes easy. They are durable and throw a soft delicate line. Today a Phoenix silk line will set you back some £40 or near NZ$120, a far cry from the 1923 price. For the modern fly fisher a disadvantage is in having to unwind the line from the reel after every outing and to prevent the dressing varnishes becoming 'sticky'. However, my good friend, Barry Dunkley, who is nothing if not innovative, buys undressed silk lines from an English firm, dresses them himself using magical brews and produces beautifully soft floating fly lines which are a pleasure to use.

Silk fly line in box.

With the advent of plastic coated fly lines, dressed silk lines are now almost obsolete. Not only are plastic lines more robust and require little maintenance they come in a wide variety of style, weight, colour, and price. Modern floating plastic lines are filled with microspheres trapped in pliable plastic coating with superb smooth surface. Most manufacturers claim plastic lines never need lubricating and need only a wipe down after use but I must be unlucky. Without exception all the fly lines I have owned after some use, decided to sink unless I apply a very *light* application of 'Mucilin'. It is now easier to choose a line that will balance the rod during a normal casting sequence thanks to a line manufacturers modern code. This matches a fly line to a particular rod by giving it the same number on a scale of 1 to 10 and relates to the weight in grains (437.5 grains = 1 ounce) of the first nine metres (30ft). The higher the number the heavier the line.

American Fishing Tackle Manufacturers Standards.

Line Number	Weights in grains	Range in grains
3	100	94-106
4	120	114-126
6	160	152-168
8	210	230-250
10	280	270-290

FLY LINE STYLES

Double Taper: This line is level over most of its length then tapers gradually for around 10 metres at both ends. This feature affords smooth casting and a soft delivery of the fly or nymph. The double taper of course allows the line to be reversed when one end becomes worn. For general river trout fishing it is the most popular of styles. It is also economical. If you feel enterprising and brave enough, you can cut the fly line in half and attach each section to a backing of 20lb nylon monofilament. Two for one.

Level: Level lines are still used for wet fly or lure fishing but they are not so versatile nor do they deliver the fly with the same finesse as the double taper. But occasionally I meet fly fishers using not only level lines but level leaders into the bargain and marvel at their skill. This is worth noting in that although good tackle is important it will never replace streamcraft or intuitive angling skills.

Weight Forward: These lines have a shorter taper that increases in diameter towards the tip, the main section of the line being level. This forward weight is of great advantage when 'shooting' the line to gain longer distance but they lack the sensitive delivery of the double taper.

Sinking: Having a greater specific gravity than water these lines allow the line to sink to a required level and are particularly useful in lake fishing. The smaller the line weight the slower the line sinks.

Sink Tip: This line allows only the tip to sink while the main body of the line floats. Very useful in helping the nymph down to the trout's level and with most of the cast floating makes the back recovery cast easier.

A fly line, unless flogged to death, should give good service over two or even three seasons without *undue* cracking. Even then it can be pressed into service as a wet fly line and, as the wear usually occurs on the first seven to ten metres, it can even be used as a sink tip. A word of advice: If the rod tip is ceramic, as against steel, occasionally check the inlaid ring. A chip or crack will very quickly ruin your fly line.

Fly lines come in many colours. Some argue that a light colour floating fly line, when viewed from below, is less obvious to the trout than a dark one. This is debatable because, in my opinion, if the trout sees the line, whatever colour, you might as well go home. The only part of the fishing outfit that should be near the trout is the tippet or the fine end of the leader and that, if fine enough, should be practically invisible. So does colour matter? The answer will be clear enough if you overcast near a trout lying in an avenue of willows. When cast against such a dark background a white or pink line gives off a definite 'flash' and can easily frighten fish. On wide rivers or when fishing for deep lying or insensitive fish, bright lines are no problem although I fail to see why such fancy colours are used. If nymphing, surely a small sighter is the answer in keeping track of the business end. As usual, it comes down to common sense.

Almost there!

4

Leaders and Knots

Leaders

The leader provides the important link between the fly line and the fly. It must deliver the fly to the trout as naturally as possible and can be constructed with sections of different strength nylon monofilament or bought in knotless tapered lengths. In both cases the leader should turn over smoothly. Some nylon is stiff, some relatively soft. Good turnover leaders have a butt and middle section of the stiffer nylon and terminate in a long tippet of more pliable nylon that allows the fly or nymph to settle without alarming the trout. A suitable knotted leader will have these proportions: Butt 60%, Taper 20%, Tippet 20%.

Leader length depends on fishing conditions. For low or still water conditions with alert trout a leader of 12-16ft (3.6-4.8m) may be required whereas an 8-10ft (2.4-3m) leader will be sufficient for a more disturbed surface. Shop-bought solid-tapered leaders are usually around 2.7m (9ft) in length but with a suitable X rating they can be extended with tippet from a spool of finer nylon. Why an X rating? If you subtract the X rating, say 4X from 11 you find number 7 and you then know that the tippet diameter is .007 which has a breaking strain of around 6lbs. For 6X you find .005 and a b.s. of 3lb. Breaking strains and diameters vary with the brand of nylon.

Some tapered leader specifications:

	1X	2X	3X	4X	5X	6X	7X	8X
Tip diameter	.010	.009	.008	.007	.006	.005	.004	.003
Butt diameter	.021	.021	.021	.021	.021	.017	.015	.013
Test Kg	4.1	3.18	2.7	2.2	1.8	1.4	0.9	0.45
lbs	9	7	6	5	4	3	2	1
Hook size	4-8	6-10	10-14	12-16	14-22	16-24	18-28	18-28

Knotted Taper

Some fly fishers believe that knots in a leader lead to tangles but if the leader is constructed properly it should give no problems. The advantages of knotted leaders are that after an initial outlay they will cost much less than the commercial variety and they can be made up to suit a variety of fishing situations. The foremost requirement for 'made-up' leaders is the ability to tie neat and very secure knots. (See later section). A well-designed leader conforms to the 60-20-20 formula which is 60% butt or stiff section, 20% made up of short similar sections and 20% single strand (tippet) of softer nylon. For efficient matching of leader to rod and line, experience should be the teacher but the following specifications should give the beginner a flying start:

4.5 metre (14.5ft) leader with 5X tippet for hook size 14 to 22.

Length			**Diameter**		
mm	inches	B.S. lbs	mm	inches	X
1100	44	20	0.42	0.17	—
850	34	15	0.37	0.15	0
450	18	12	0.30	0.12	1
400	16	10	0.22	0.009	2
,,	,,	,,	,,	,,	,,
,,	,,	,,	,,	,,	,,
625	25	6	0.17	0.007	4
375	15	4	0.12	0.006	5

2.7 metre (9.0ft) leader with 5X tippet for hook size 14 to 22
Butt- taper - tippet diameters the same as for the 4.5 metre leader, only the strand lengths changed.

mm	in	mm	in	mm	in
700	28	250	10	150	6
350	14	150	6	150	6
300	12	150	6	500	20

It is most important to relate the hook size to the tippet diameter. A small fly tied to a stout tippet will result in poor presentation.

X	BSlbs	Fly Size	X	BSlbs	Fly Size
6	3	16-18-20	3	6	10-12-14
5	4	14-16-18	2	7	8-10-12
4	5	12-14-16	1	9	6-8-10

Butt sections may be altered to suit the fly line. The above leaders will match # 5-6-7- fly line tips. In general, butts should be ⅔ the diameter of the fly line. The above shorter leader is recommended for beginners.

KNOTS
The ability to tie good knots comes only with practice and plenty of it and a large fighting trout is one of the best teachers. After a hectic battle how many of us have been left gazing ruefully at the tiny flyless pigtail. There does come a time however when a fly fisher becomes very proficient with knots and with a big fish on the end of the line, in a tricky situation, he or she needs all the confidence possible.

Clinch Knot (Nylon to hook)

The most common of all angling knots — the clinch — is easy to tie. Pass the nylon through the hook eye and wrap it around the main stem five turns, then pass it back through the opening between the first wrap and the hook eye. Snug the loops up by pinching the knot and pulling snug. An improved version can be made if, before the knot is tightened, the nylon end is taken back up the main stem and passed through the upper loop.

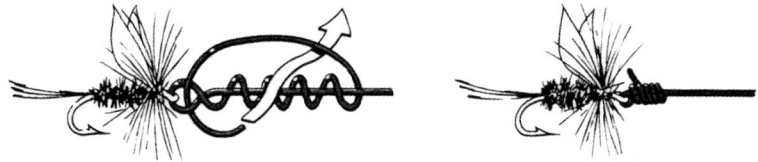

Needle Knot (Leader to line)

There is no neater knot for this purpose and if tied properly it will never fail. While the knot can be tied at the streamside the workbench is by far the best place. I recommend the use of a long hatpin, tweezers, a small pair of pliers and a vice. With scissors cut a point on the butt nylon. Place the pin in the vice and lay the fly line against it with say 15mm spare beyond the point. With say 100mm spare of the butt nylon, pinch it against the nail and the line and wind five tight neat wraps back down towards the point. Loosen slightly and after taking the butt tip in the tweezers feed it back under the five wraps. You will bless the nylon point. Now grip the exposed, short, nylon butt in the pliers and gently ease the needle out from the wraps at the same time pulling the nylon butt with the pliers. As the needle becomes completely free, tighten hard. This chokes the line and it remains only to angle cut the spare line and nylon ends. Do not hesitate to cut well up to the knot. Again, if the knot has been tied correctly,

there is no fear of the leader ever pulling free.

Another but similar method can be used at the streamside by using the eye of a blunt, stout, 5cm needle to thread the nylon back under its own turns. The potentially dangerous point should be broken off before use. The only difference between this and the preceding method is that the tightening of the nylon coils is hidden, but with practice very neat turns result.

Blood Knot (Nylon to nylon)

The standard knot for joining two pieces of nylon, particularly for constructing leaders. In tying this knot I use a little trick which with practice makes tying this knot a doddle. One word of warning: joining nylon of significantly different diameters is unwise as the knot will be unreliable. Cross two sections of nylon with 500mm free tags and holding the X with the left thumb and forefinger TWIRL four turns of one tag to the right keeping the main nylon stretched in the right palm. The right thumb and forefinger now take over from the left which TWIRLS four turns to the left keeping the main nylon stretched in the left palm.

Now the trick. With thumb and forefinger of both hands, PUSH the turns on both sides of the X to the centre and 'presto', a small loop will appear. Right thumb and forefinger now picks up the left tag and after holding the X, threads it through the loop. The left thumb and forefinger pick up the right tag and thread it through the loop, IN THE OPPOSITE DIRECTION, then press X (and both contained tags) while the right thumb and forefinger pulls the tags. At this stage the loose knot can be exposed while the knot is tightened. It never fails to fascinate as the left and right coils slide together in one of the neatest knots imaginable.

Surgeon's Knot (Nylon to nylon)

If you get caught out in poor light with a broken tippet and the trout are rising all around you then tying a blood knot can be a problem. Not so with the surgeon's knot. Simple to tie and very secure its only disadvantage is a bit of waste nylon. This is because you need at least one end at least 600mm to form a loop sufficiently large to tie the knot. No tricks to this one, just cross the ends, form the loop and pass one pair of ends twice through and around it. Whereas in the blood knot one of the waste ends make an ideal dropper for the wet fly, the ends of a surgeon's knot appear at an angle and are less suitable.

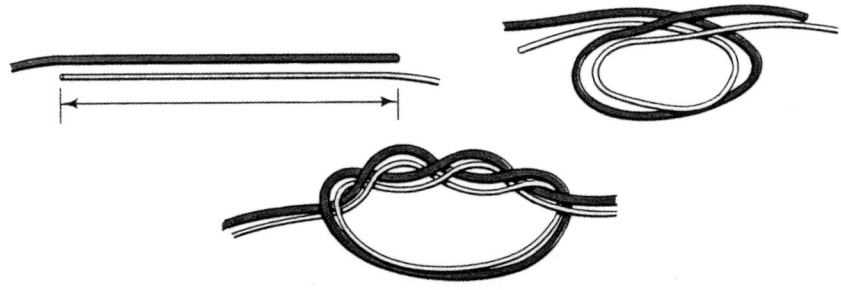

Backing Knot (Backing to line)

The neatest and smoothest way to join backing to a fly line is to spin a fly tyer's bobbin loaded with thread around them then secure with half hitches. In the happy event that a very large fish has run you to the backing, this smooth joint will not snag in the top ring. For non-tyers (hopefully less after reading this book) a neat backing knot is recommended. With the backing in the left hand form a loop some 120mm from the end and lay the near end of the fly line against the bottom of the loop. Bring the backing end under the fly line and through the loop towards you, pinching the tiny coil with left thumb and forefinger. Repeat this for five turns, keeping the coils butted but not overtight. Hold the knot firm and pull the main backing line to tighten. This snugs the knot and both tag ends can then be trimmed. Take care to leave at least 5mm of backing tag exposed but the line can be trimmed close.

Spindle Knot (Backing to reel)
An easy knot to tie. Simply pass the backing end around the spindle and tie an overhand knot on the short end. Tie another overhand knot between the spindle and the first knot and pull smoothly until both knots lay snugly against the spindle.

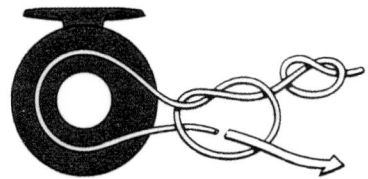

Pear Knot (Nylon loop)
Many anglers prefer some easy method of changing either leaders or tippets, using nylon sections with looped ends. The pear knot allows this and as the name suggests gives a perfect shape instead of a cocked noose. Tie an overhand knot 60mm from the end of the nylon leaving it open, then pass the end back through and tighten. Tie another overhand knot as close to the loop as possible and draw the two knots together.

Despite everything there will come a time when you come across a fish that will defy almost any knot. I recall where a high stopbank looked an ideal place to spot a fish. And there he was — as long as a man's arm, apart from a lazy wave of fan size tail, he lay almost motionless in the little side stream. It flowed through the long green pool with just enough current to make him seem almost motionless. I backed the van a metre or so to where Jean could see him out the van window. 'He's a big one!' said Jean. Well, he certainly frightened me. Even the sight of him made me nervous but faint heart never caught a trophy so, after moving quietly away, I tackled

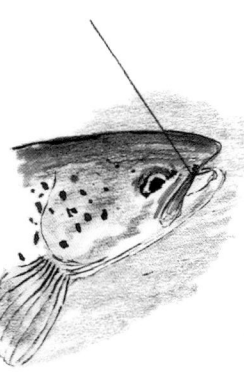

up and returned, heart thumping and knees knocking. He was still there, floating midwater.

I knew that my chances, in that almost still, clear, water were slim indeed. The big decision was what to tempt him with — a tiny nymph or an equally small dry? I plumped for the former after deciding that any surface crinkle would alert him and hoping that, if dozing, a nearby 'plink' would catch his attention. It did and as he gently moved forward towards the #16 brown stonefly nymph I tensed in readiness. I saw the white of his jaws.

What happened after that was a blur as, after feeling the little hook bite, he tore off downstream out of the backwater. It was an unequal battle. As I tried to scramble over the giant stopbank boulders trying to keep up with the speeding fish I heard Jean shouting encouragement. I needed more than that. What I really needed was a miracle. In the far distance, now in the fast flow of the river, I saw him leap a couple of times and as the final turns of the backing appeared, resigned myself. For a surprisingly long time the hook held but then came the dreaded slack. Naturally disappointed, I reeled in the flyless leader and tippet. After climbing back to the van, I saw the funny side. I could just imagine that trout's comment: 'What a damned cheek!'

Ready to take off!

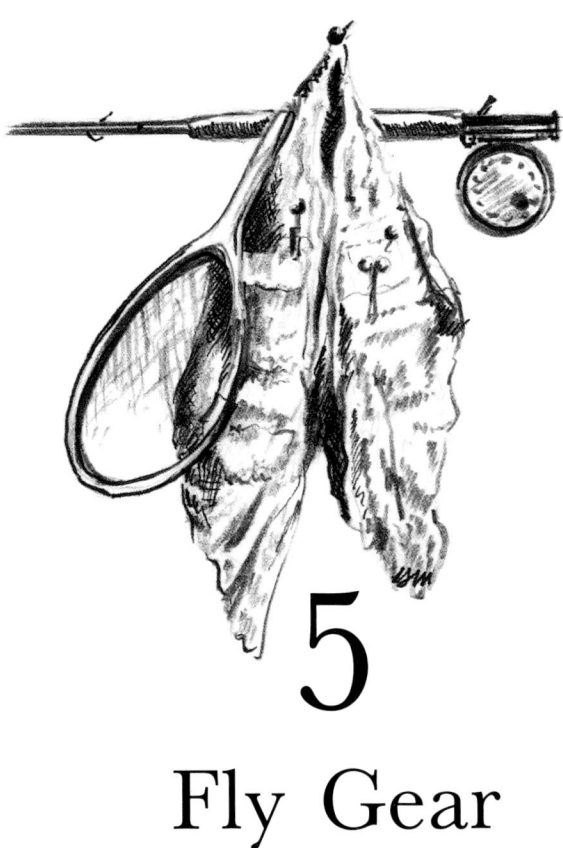

5

Fly Gear

Fly Reels

Although a reel can be considered as merely a line storage item it has a more important function. When, after much effort, cunning, and often perseverance, the hook goes home and an angry brown or a racehorse rainbow screams across the pool, the only thing between failure and success is a smooth running reel. This doesn't mean you have to rush out and spend a small fortune on a top drawer reel. A very satisfactory reel can be bought for a quarter of the price of those aristocrats and still give good service. Most reels have a simple mechanism and it is rare for them to jam during panic stations but a well-designed drag system, one which can be easily and quickly adjusted is important.

Line capacity should be for 30 metres of Double Taper floating fly line (DT6F) and 100 metres of Dacron or Micron backing. If a new chum, the choice of left or right hand wind is important. For maximum efficiency the reel handle should be on the line hand side which allows playing a fish without swapping the rod over. Never having had a tutor I have spent many years doing it the wrong way, but old habits die hard. Durable and smooth line guides are essential otherwise the fly line will quickly be ruined. From experience, I recommend that you disfigure the new reel by scratching your

Left. Brass Malloch Pike and Salmon spinning reel. Reversible spool. circa 1920.
Middle. Hardy's salmon reel circa 1930.
Right. Boy's brass reel circa 1930.

address on it. Sooner or later you will bless me for this advice and especially if it is lost while still attached to a rod!

Wading Gear

I rarely quote other writers but the following piece is so apt to the subject that I can't resist. In his *Days and Nights of Salmon Fishing in the Tweed* (1843), William Scrope wrote the following on wading:

> When you are wading through the rapids, step on quickly and boldly, and do not gaze down on the stream after the fashion of Narcissus; for running waves will not reflect your beauty, but only make your head giddy. If you stop for a moment, place your legs abreast of each other: should you fancy a straddle, with one of them in advance, the action of the water will operate upon both, trip you up, and carry you out to sea.

Even better:

> Avoid standing upon rocking stones, for obvious reasons; and never go in water deeper than the fifth button of your waistcoat; even this does not always agree with tender constitutions in frosty weather. As you are likely not to take a just estimate of the cold in the excitement of the sport, should you be of delicate temperament, and be wading in the month of February, when it may chance to freeze very hard, pull

down your stockings, and examine your legs. Should they be black, or even purple, it might, perhaps, be as well to get on dry land; but if they are only rosy-red, you may continue to enjoy the water, if it so pleases you.

They don't make 'em like that these days! In New Zealand many anglers wade 'wet' during the warmer months resorting to waders only during spring and autumn. While a pair of old gym shoes with glued-on carpet soles are light to wear they wear out quickly. Unfortunately, proper felt wading boots are pretty expensive at around the price of a low range graphite or a high priced reel. But if the pocket can stand it they are one of the best investments you can make. What price being safe and comfortable? It pays not to let the soles wear too thin before sticking on replacement felt. Thick industrial felt is suitable and combined with ADOS F2 adhesive a pair of boots should last a few seasons. If you want to wade dry and use wading boots you have to buy a pair of wading stockings. These are lightweight, seamless and pack into a very small area. But if you break the cost down over a few seasons they are not as expensive as they seem. 'Aquaseal' is excellent for patching and should be kept frozen when not in use.

Rubber-soled waders can be treacherous on slippery algae but these are the traditional anglers' footgear. It is possible to have the soles roughened and felt applied but it is a messy business. Felt-soled, nylon fabric thigh waders are a boon to the angler who rarely fishes deep. They are easy to slip on and off, and when nature calls they are more convenient than chest

Nylon stocking chest waders

Canvas booted waders

Felt soled boots

Nylon thigh boot waders

Thick socks

Nylon stocking waders in pack

Wading sneakers

waders. Neoprene waders are also expensive and although lovely to wear in the chilly waters of spring they will cook you if the weather turns warm.

Landing Nets

Some fly fishers consider nets a damned nuisance yet others feel naked without them. Only experience will tell you to which camp you belong. The argument that they get in your way can be solved by using a net scabbard slung across the back. One thing is certain, at some stage there *will* come a time when, despite every effort, a very desirable trout will be lost through lack of a net. Tear drop nets can be a nuisance especially when pushing through scrub or bush but many anglers tolerate them because they are so handy. If you do plump for a tear drop net make sure the opening is large enough to let a six-pound (2.7kg) trout in without arguing about it. The collapsible, extendable vee net is my choice, safely nestled in a scabbard. I have had a few over the years and find them very good. The present net is over 10 years old, has had plenty of wear (thank goodness) and is still in fair condition. Using a net also makes it easier to hold a fish in midstream and release it relatively unharmed, a consideration for conservation-minded fly fishers.

Fly Fishing Vests

These days it is uncommon to see a fishing bag. In the past every self-respecting angler toted one. They ranged from fancy shop-bought items to the almost traditional army surplus gas mask bags. I always remember fishing bags as something to give you an aching shoulder and, after carrying a fish or two, it was best left to hang outside a shed. For some years now, fishing vests have been popular and have become almost essential to the fly fisher. Unfortunately, the more pockets the more gear the angler packs, much of it unused. Lightweight nylon net vests are ideal, especially in summer. Last year my fishing mate Vern introduced me to the belt pack which allows ample room in the pouches. The advantage is having your gear handy at waist level instead of fiddling around in upstairs pockets.

Polaroids

Of all the incidental items, there is one that will increase the angler's success rate a hundredfold — Polaroid spectacles. No experienced fly fisher would go fishing without them. During bright days smoky grey lenses are good but on dull days I find the amber-tinted lenses are the most effective. Great care should be taken not to scratch the lens. For many years I have used Polaroids with a small magnifying insert below the main lens which dispenses with the need to change spectacles for close-up work. They save much valuable fishing time. They are made by 'Fisherman's Eyewear', an American company but I have yet to see them available in New Zealand.

Belt pack

Soft hat

Vest

Net
& sheath

Thick socks

Felt soled boots

Back pack

Wading stick

Fly Boxes

Fancy aluminium multi-compartment fly boxes are expensive, handy, but
not essential. If you are into status symbols a hint or two just before
Christmas time could well do the trick. High impact plastic boxes with foam
linings for nymphs and wet flies are quite satisfactory. If you get a ducking
make sure you remove and dry the contents which quickly rust. There are
many different styles but I favour the saw-tooth foam variety as against
flat foam. It matters little which you choose. Sooner or later you will lose
it and the less expensive, the quicker your tears will dry. The prudent fly
fisher strings both dry and nymph boxes to his vest.

Nylon Dispensers

Hanging on my workshop wall are a variety of these ingenious gadgets.
They range from a circular box with no less than six small spools inside
to another with six spools in decreasing diameter one inside the other, similar
to those quaint, ever-decreasing Russian dolls. Others in my collection I
find impossible to describe. I think it will be their last resting place.
Nowadays, a couple of spools of 5X and 6X nylon for tippets is all I carry.
I have learned my lesson and studiously ignore the gadget sections of fishing
magazines.

Zingers

I don't think this word is in the dictionary but it aptly describes this compact pin on spring and cord item. Some may consider zingers a luxury, making do with a piece of string, but they are so handy to attach clippers or a hook sharpener (small diamond grit file) and store them safely after use. One zinger is enough for me and when I see some of my American fishing friends with an array of them and everything but the kitchen sink dangling I wonder. A word of advice: inside these handy cases is a little spring which, once released, has a Machiavellian sense of humour. Check the small centre screw, often.

Forceps or Pliers

It makes good sense to return trout with the least possible harm. This means handling the fish as little as possible. By far the best way is to leave the trout in the net and hold it gently but firmly through the mesh while removing the hook. If the hook is barbless so much the better but otherwise a pair of artery forceps or small long-nosed pliers are the answer. Trout bounced about the beach will be damaged as will those handled by the gills. Fish found bleeding at the gills should be dispatched and for trout up to 1kg a forefinger in the mouth, a thumb on the neck, and strong applied backward pressure will snap the backbone.

1st Aid

Pliers

File

Muslin

Various

Float
powder

Dry fly box

Polaroids

Zinger
& nippers

Insect
box

Spare
nylon

Nymph box

Select
fly boxes

Belt pack

Small items but nevertheless important. According to makers of plastic fly lines they only need to be wiped to keep them afloat. As I mentioned earlier, I must be the odd man out. After some use, mine tend to sink unless some floatant is applied. A small tin of Mucilin is a handy item.

Indicators come in a variety of guises. These bright tags are a boon to the nymph angler when casting to unseen fish. A piece of fluorescent wool well rubbed with Mucilin is as good as any but it can be a nuisance if you want to remove it quickly. On the American market are packets of lanolin-impregnated yarn which makes a great floater. During his last visit, Gary Borger left me with some *black* yarn. When paired with a bright coloured sighter, whether in glare or shade, both can be easily seen. New on the local market is a bright yellow float paste which is soft to apply and hardens in the cold water. An advantage is that, unlike wool, it can be easily crumbled from the leader.

Although I'm not a great fan of lead when used as additional weight I have to admit it is necessary when fish are lying deep and they are giving me a hard time. Small lead split shot is useful, as is lead strip, but removing and replacing it is a nuisance. This problem has now been overcome with soft weight, dark putty-like material that hardens on contact with cold water and, like float paste, can be easily removed from the tippet or nymph.

Wading Sticks

It surprises me how few fly fishers use a wading staff. Apart from the safety aspect when caught in fast water they are most useful under normal wading conditions. That third leg allows the angler to give his or her undivided attention in searching for trout, especially on bouldery, algae-covered river beds. Custom-built wading staffs are available but expensive and a stout stick is easily procured. But make sure it is strong. Last season I put my trust in a willow stick which snapped at the wrong moment and I tumbled down the rapids. Not a very pleasant experience and very expensive — it cost me a Loomis fishing rod and reel. If age brings wisdom I am still waiting.

To the great unwashed, we anglers, encumbered as we are with all our fishing gear, are often seen as ridiculous. But I always remember the story of the horse that, for lack of a nail, lost a shoe, and threw King Arthur who lost the battle. Or was it General Custer?

6

Fly Tying Materials

If there is an Oscar for collectors then I'm sure that stamp or coin buffs would be in front of the queue. But close on their heels would be the fly tyers. Whether he is taking the offspring for a walk around the park, and ogling the peacocks, or on a Sunday drive, keeping an eye open for some unfortunate hare, the dedicated fly tyer is always on the lookout. A human magpie. Most of my collection, once seeming treasures, rarely see the light of day. The fly tyer, always creative, harbours the hope that one day he will produce a 'superfly', a pattern that will earn him a place among legends.

Describing an Aladdin's Cave of fly tying materials is a formidable task but probably the best place to start is with feathers. They come in hundreds of shapes and colours, indispensable for imitating trout food insects. It *is* possible to tie a dry fly with fur or hair, for example, Hares Ear, or Humpy, but for the majority of floating flies the feather reigns supreme. Close on the heels of the feather is fur. Again, the variety is almost endless, ranging from polecat to polar bear and from rabbit to racoon but hare or opossum fur is probably the most popular. For body materials the list is endless with peacock and ostrich herl, floss silk, chenille, tinsel, wool, and other natural fibres. Synthetic fibres such as polypropylene and swannundaze and microfibbets for wings, tails, and body dubbing are now readily available. Some superior to natural material. But despite the wealth of fly tying materials a beginner may well make do with the following items:

A BASIC LIST OF FLY TYING MATERIALS

1. A cock cape. Try not to economise with this. A cape will provide hundreds of usable hackles. Most useful colour — dark ginger.
2. A hen cape. These are not expensive. A black and a brown are recommended.
3. A rabbit skin, opossum skin, dark grey with red/brown belly fur, hare skin, and deer hair patch — red and whitetail.
4. Peacock eye feather, partridge plumage, cock pheasant tail.
5. Tying thread (waxed), floss silk, gold tinsel and thread, copper and lead wire.

   The above basic items will cover well over 100 patterns.

FLY TYING MATERIALS

Feathers (Hackles)

The main source of hackles for dry flies are rooster feathers taken from the birds' neck crest around the throat. By skinning, instead of plucking, the hackles stay in a convenient package. Loose hackles are a nuisance. Cock hackle quality varies greatly from semi-wild Asian roosters with variable quality hackles to the long, short fibre hackles of the genetically bred (but expensive) domestic rooster. Less expensive capes still carry reasonable quality feathers but (except for small flies) may require two hackles per fly. American domestic capes come in three grades and need one hackle only. To most tyers the cost of domestic capes seems high but based on the two to one hackle argument they can well be economic. I compromise, treating myself to a top grade cape now and again but keeping my eyes open for Asian cape bargains.

They may not be in the same class but they still produce a good quality trout fly, providing care is taken in selecting the cape. One of my favourite day dreams is to be wandering around some exotic backroads of Thailand (I have seen some great capes strutting around some of the Mayo tribes villages) with no other purpose than to point out a splendid Bronze Grizzly or a rich red Coch-y-bondhu or Furnace. Then for $20 (American of course!) have a dozen top quality capes delivered to my hotel and drooled over during our 'happy hour'. Sometimes our dreams come true but, because of agriculture import restrictions, not in this case. Many years ago I wrote of a visit to a Southland farm. At some stage I noticed a nice cock bird scratching in the dirt behind a shed. Another appeared together with a bevy of hens. Both roosters were old with spurs like scimitars and even a glance told me they owned top notch capes. Business done I steered the conversation to trout fishing, and made the first cast.

Peacock 'eye'

Pheasant
Tail

Pheasant
(hen)

Grouse

Mallard
'blue'

Mallard
Quill

Goose
Quill
(dyed)

Pheasant
(breast)

Pheasant
tippet

'If you have any old roosters that kick the bucket I'd be glad to have them,' I said casually.

'How many do you want?' he replied and, before I could answer, he was on his way into the house. When he reappeared it was with a double-barrelled shotgun.

I don't know how they do it but fowls seem to have a sixth sense when a shotgun appears. They had already got into gear when the first blast not only took a slow rooster's tail feathers off but a corner of the shed as well. By this time the mixed flock had really got the message, some running around in tight circles while others were mere specks on the horizon.

'B----r em,' said the excited farmer. 'We'll give 'em a rest.' But that was short lived as one of the prime roosters, after poking his head around the corner, panicked and half flew, half scuttled, across the yard.

'Better than shooting ducks,' he chortled as the bird shot sideways about two metres and, for a few moments, was hidden in a cloud of feathers.

That was my first taste of rooster hunting and I wondered if there was anything left of the bird. But, although there were plenty of missing feathers, its neck was untouched.

'Now for the other b----r!' he said, starting off towards some bare-branched trees where some of the birds were saying their prayers. Now rather deaf, I quickly indicated that one bird would do quite nicely, thank you, and took my leave.

"Come back sometime for the other one!' he called after me as I left clutching the prize. I never did.

DRY FLY QUALITY HACKLES

A prerequisite for a dry fly hackle is that the fibres should be reasonably stiff, bright, and short in proportion to quill length. Comparing Asian and domestic hackles for tying a #12 dry fly, the former may have a quill length of 50mm and fibre length of 10mm whereas a Metz hackle may have a quill length of 70mm and a fibre length of 8mm. This may not seem a great difference but because Asian hackles taper sharply, only a fraction of the fibres rest on the stream surface. The length of the genetically bred hackle fibres is consistent for 90% of quill length giving the fly far superior buoyancy. Apart from this advantage they are much superior in fibre strength.

SELECTING CAPES

Before buying a cape it should be checked for the following features:

1. Consistent colour and shine. While the rear of a cape will be less colourful and brilliant, capes with too marked a difference between back and front should be avoided. A *small* amount of 'web' (softer fibres near the butt) is no disadvantage in that it helps to back up the stronger fibres.

2. Resilience of the fibres. When a hackle is stroked against the natural lay it should spring readily back to normal.

3. Minimum down at quill butts.

4. Short fibres in relation to length.

5. Full complement of the smaller hackle range.

HACKLE COLOURS AND PATTERNS

Hackles come in quite a variety, certain trout fly patterns requiring a certain hackle. The hackle feathers listed below are a fair example (the most popular hackle types are indicated *):

Ginger*, Ginger Grizzly, Red, Dark Brown*, Red Furnace*, Brown Furnace, Badger, Pale Badger, Grizzly*, Cream Grizzly, Blue Dun, Black*

Hackles can be readily dyed using domestic dyes. Red or dark ginger capes are suitbale for dying black and white or cream capes for dying blue/grey. Grizzly capes can be dyed a dark mahogany (my favourite colour). Hot water dyes are recommended but take care not to boil the cape as, on drying, the skin will be over-brittle. Remember to add a little vinegar to the water to fix the colour. Before dyeing, always thoroughly wash the cape in a detergent to remove natural grease. After dyeing, dry in sunlight or a superheater cupboard. After removing the cape the remaining solution can be used to dye fur. A purloined old saucepan is ideal together with a wooden spoon. If using the domestic range or cooker make sure there are plenty of tissues or newspaper to hand otherwise you may outstay your welcome.

HEN CAPES

Hen hackle fibres are soft and water absorbent, which makes them most suitable for wet flies which should sink quickly. For every old rooster there are thousands of hens destined for the table and that makes hen capes much easier to obtain. Selecting colour is really the only requirement, although finding the more uncommon capes is not easy. As with the rooster capes, they should be stored in plastic bags together with a small tablet of napthalene, an effective moth repellent.

PLUMAGE

Plumage can be described as a feather from any part of a bird other than the rooster or hen. It is safe to say that at some time or other, someone, somewhere, will have used plumage from just about every bird in the world. There are some who swear that one or maybe three trout fly patterns will catch just as many trout during the season and scorn the packed fly box of others. These anglers are to be admired for their fortitude and resolve. Not many of us want to risk it. We not only want to catch trout but we want to have some fun as well and when, after much teeth grinding and

Fly Tying Materials

1. Heron herl
2. Grouse
3. Pheasant
4. Black hen
5. Red Asian hackle
6. Grizzly American
7. Blue Jay
8. Partridge
9. Deer Hair
10. Opossum fur
11. Chennille
12. Floss Silk
13. Peacock 'eye'
14. Silver flat
15. Gold flat
16. Mallard quills

Acrylic yarn

Speckled
Brown Hen

Poly yarn

Calf Tail

High Vis

Duck Quill
(Peverill)

Mohair

Wool

Chennille

Opossum fur
(belly)

Mallard
(breast)

Latex

Partridge
(dyed)

Deer Hair

Micro fibbets

fly selecting, we finally come up with the right one it's like winning a lottery or solving a Ngaio Marsh mystery.

A plethora of plumage. From the tiny flank feather of the starling to the great tail quills of the turkey; from the back feather of the partridge to the glorious 'eye' feather of the peacock and much more — they are all essential for some particular pattern. Some have achieved such notoriety as 'trout killers' that their use has been banned as in the case of the bittern. The latest entrant for honours is the curly 'oil' feather from a mallard's bum and titled Cul de Canard. Despite making fun of them they are very effective for tying emergers.

PLUMAGE (Detail)

1. Peacock Herl: When my neighbour arrived bearing one of his peacocks, savaged by a dog, I promised to church-go the next Sunday. A full peacock skin is a sight to behold, a feathery bundle of rainbows. No words can describe the brilliant emerald back feathers or that lustrous ultramarine collar. It is still there in the box after many years with the only part used; the tail feathers. This is what drives the tyer to new patterns. He secures some odd but delightful plumage and, by hell or high water, he's going to develop a winner out of them.

From the 'eye' feather comes the 'herl', the bronze thicker fibres just below the 'eye'. It is used, as an example, for the bodies of chunky trout flies such as the Coch-y-bondhu and Royal Wulff. Only the well-fibred strands should be chosen. The 'eye' fibres are used for quill bodies after stripping away what is known as the 'flue' leaving a fine striped quill. All 'eyes' should be checked for this feature by bending the whole 'eye' tightly over the fingers. The paler the quill the better will be the stripe when wound on the hook.

2. Duck: For dry flies, duck wing feathers are very popular both for the wings of a trout fly, wet and dry, and the wing cases of nymphs. For dry fly wings the primary feathers of the mallard still stand supreme. The quill fibres have that intrinsic ability to 'lock' as in a zip fastener, a feature which helps maintain wing shape.

Now we have one of the inconsistencies of trout fly imitations. A mayfly, afloat, holds its wings erect and folded. The traditional mayfly imitation is always tied with the wings slightly spread. This, supposedly for the aerodynamics in the hope that the fly will settle upright. Fortunately, as long as the fly looks good to eat such finer points seem academic. But tradition dies hard. I'm quite sure that although trout have super eyesight when it comes to spotting anglers they don't get a very clear picture of trout flies from their subsurface viewpoint. For nymph wing cases both the grey from the primaries and the secondaries are used as are the vivid green/blue

secondaries of mallard drake. The former is tied with the shiny side top as is the blue feather. Bronze breast feathers are a useful substitute for the American Wood Duck.

3. Pheasant: While some traditional patterns demand hen pheasant tails it is the cock bird that is the most useful. Russet/red cock pheasant tail feathers are essential for tying the popular Pheasant Tail nymph. They vary considerably in fibre length which should be preferably at least 50mm. On smaller nymphs this is sufficient to tie tails, body, and the wing case of the artificial. Pheasant body plumage is used for many lure patterns including 'church window' back feathers for Mrs Simpsons and my pattern of the passion fruit hopper.

4. Partridge: Good quality brown partridge back feathers are not easy to obtain but are well worth the search. Much of the imported feathers, when packaged, contain a liberal proportion of large feathers much less useful than small ones. Partridge hackles should measure around 15mm across or smaller. They should be well mottled and brown, not grey although these are mixed when packaged. Short fibre feathers from grouse and partridge are few and far between but this can be overcome by using the 'wrap hackle' technique (shown later).

5. Turkey: These very large dark umber turkey tail feathers make splendid nymphs. The fibres are extremely long (50-60mm) giving plenty of length for tail, body, and wing case of the artificial, such as the Turkey Tail.

6. Grouse: Similar to the partridge but a much darker feather and used in many traditional wet fly patterns and lures.

7. Starling: One of the most effective feathers for emergers, that stage of interim mayfly development between nymph and adult. Perfect in reproducing that battered appearance of aborted, helpless sub-duns. Provides not only a soft hackle effect but a seductive sheen.

FUR

We humans have long discarded the need for fur, our substitute being a cosy fire or an electric blanket. Fly tyers and fashion designers are exceptions. Fur's use as a trout-fly tying material goes back a few generations. One of the early references is in Izaak Walton's *Compleat Angler*. In his list of 20 recommended patterns nearly every one contains a fur dubbing. While the list highlights the effectiveness of fur, the fanciful varieties show Walton as a theatrical. Here are some of the furs he mentions: Hogs ear, black, long coated cur, (note, not pedigree), bears hair, brown brended cow (note,

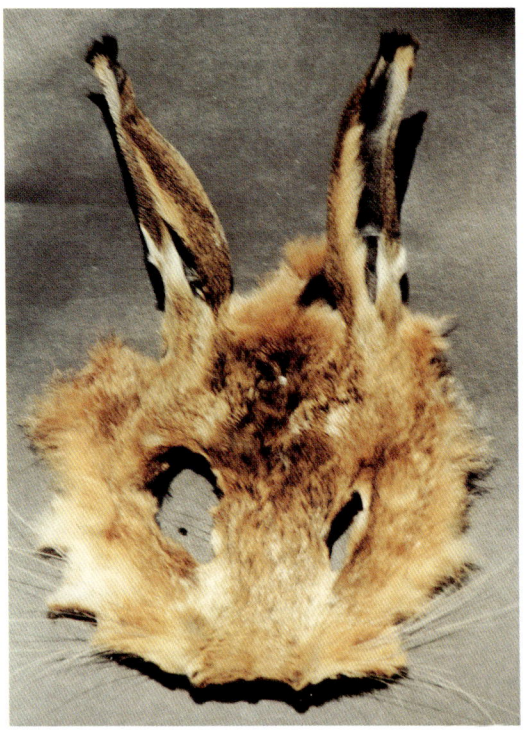

Hare's mask.

brended!), bottom of squirrel tail (note, not top), camel hair roots (camel hair available in England in the 17th century?), black greyhound, black water dog (waterdog?), hair of abortive calf (note, not your everyday material!), fox cub, sable, fitch, white weasel, badger and sanded hog.

Today's tyer can make do with four types of fur or hair:

1. Hare fur: From light fawn or ginger through to black/blue, the hare provides a premium fly tying fur. Under fur dubbing with a little mix of guard hairs make an excellent nymph thorax. Also, but to a lesser extent, rabbit.

2. Opossum fur: Regarded as a pest, this leaf muncher is a welcome addition to the fly tyer's arsenal. The fur is water absorbent (ideal for nymphs) and mixed with sparkle-yarn it makes a good dry-fly body dubbing. The fox-red belly fur is particularly useful.

3. Deer hair: Many fly patterns use deer hair for both wet and dry flies including nymphs. Red deer hair is coarse compared to that of fallow or whitetail. A swatch of all three should be obtained using the red for larger flies and the whitetail for smaller floaters such as the Humpy. Deer hair is supposed to be hollow. I have never confirmed this but it certainly makes for wonderful floaters.

OTHER FLY TYING MATERIALS

Most of these materials are synthetic and over recent years they have become a welcome addition for the fly tyer:

1. Wool: Still a popular body material, used mainly on the larger flies or lures. Very easy to handle and available in fluorescent colours.

2. Floss silk: Many patterns call for floss silk. It forms a shiny smooth body part but is easily frayed. A tinsel or wire ribbing is recommended.

3. Poly-yarn: A synthetic material with a sparkling quality and effective when mixed with a softer natural fur. Also a great modern winging material.

4. Swannundaze: Don't ask me where the name comes from. I do know it makes very realistic nymph bodies, particularly stone fly imitations. A plastic string, it can be stretched to the desired thickness using hot water. When wrapped over an underbody it makes a very realistic segmented abdomen.

5. Micro-fibbets and Hi Vis: Thank goodness for science. L & L products of New York make these items. Hi Vis is a high visibility synthetic yarn for both winging floaters and dubbing for nymphs. Easy to work with, it comes in over 40 colours. Micro fibbets are the answer to fly tyers' prayers when it comes to dry fly tails. The resilient fibres are carded and glued at the butts so they are all the same length. Superior to natural cock fibres.

6. Tinsel, wire and fine gold thread: A supply of fine flat tinsel is necessary. Lurex, gold or silver, can be purchased at any sewing store. It comes in 68-metre spools, 1mm width, so you should still be using it 100 years from now. Also buy a spool of fine gold #20 thread and a #6 and #8 bobbin of brown and black waxed tying thread. These, with a spool of fine lead wire, are enough to make a start.

7. Latex: This is commonplace material but much underrated. Children's balloons provide a ready source or it can be obtained in thin opaque sheets. When cut in a narrow band and wrapped around a silk underbody it produces a very lifelike caddis imitation and, if it is thick enough, a pronounced segmented effect. Fine rubber bands are also useful.

It is surprising how fly tying material comes to hand. It doesn't matter what kind of feather comes to his or her attention, a fly tyer will always give it the once over. Will it be that elusive scrap that trout go mad for? It hardly ever works that way but when Jean said there was a dead bird by the garden shed I couldn't resist having a look. The young thrush, plumage still dark, looked a pitiful sight as he lay beside the fragments of his last and fatal meal, a snail, the shell in fragments. He'd obviously choked on some of the brittle carapace. I had the spade in my hand when I noticed the lovely soft cape feather, a subtle dark umber with grey downy fibres attached.

Before long I had some #16s tied up in emerger style and by mid-morning one of them floated towards a good trout picking *Deleatidium* duns off the feedline. On a tippet of 5x or just under 3lb there was every chance. When the smallest of dimples occurred I struck and was into him but only until he'd made it across the river and into some far bank rubbish. The next fish, only ten metres upstream was also hooked.

Having had a spell of good fishing over the previous day I decided to experiment. Fish were still active with back fins just breaking the surface, obviously feeding on emergers or battered nymphs. This time I used 4x for the tippet, the increased thickness, to my eyes, hardly noticeable. But the trout thought differently. Time and again the little brown fly was ignored and by the time I decided to return to the 5x the rise was over. A few days later at the same feedline, the river stained after rain, rising trout paid little heed to nylon size and took the same emerger readily. Apart from a couple of nice trout I learned something. Choosing the right size of nylon for a tippet is very important. In low water conditions the trout has every opportunity to closely inspect the fly and more importantly the nylon which, enlarged at that close range, may well seem like a ship's hawser. However fine, a nylon tippet must cause the fly to act unnaturally. Success comes when perhaps in one out of ten casts the fly lands just right, free of drag and directly upstream of a rising fish. One in twenty in my case.

7

Fly Tying Techniques

Whether you have the fingers of a surgeon or a lumberjack you can tie trout flies. They may be rough, or works of art, but as a beginner you will be pleased to learn that it is the former that usually catch fish! That, from the start, should give you confidence. The worst thing a beginner can do is to watch an expert tyer creating a wonderful imitation. It's like listening to Yehudi Menhuin during your first violin lesson. Far better to ask the tyer to slowly wrap a feather around a hook. Fly tying can then be seen for what it is, simple. Even the most basic of trout flies will, given the right circumstances, catch a trout. For sure you can expand on this by adding more items such as a tail or wings but rest assured fly tying is not difficult.

It is possible to tie a trout or salmon fly using only the fingers and I can only admire the skill and dexterity required but most, if not all, of those tricky fingered gentry have long departed for some heavenly fly box. Admittedly, the finest fly tying items you possess are fingers, but a few specialised tools, starting with a **vice**, makes tying, not only easier but more enjoyable. It doesn't need to be expensive. Unless you intend to supply the South Island with trout flies, a moderately priced Sunrise vice will keep you in flies for a very, very long time. I bought two of these vices in 1968 and one of them has never been used. The requirements of a good vice are that it should grip the hook without excessive force and continue to hold it without slipping. Any other factors are superfluous and/or a luxury.

A thread dispenser is just another fancy name for a **bobbin holder.** Whatever the name it is, in my view, essential for the fly tyer. When I first started tying trout flies, over 50 years ago, I had never heard of bobbin holders. I managed well enough, but when I discovered them it made the difference between flying a Tiger Moth and a Concorde. Not only do bobbin holders dispense thread without fuss they also allow some fine manipulations, poking in where fingers cannot reach, especially around the hook eye and between wings. If you entertain the luxury of two holders, make sure they are both compatible with the bobbin size and that the tube mouth is perfectly smooth.

How on earth early fly tyers managed without **hackle pliers** is beyond me. It can be tricky enough even with pliers and I find them indispensable. Hackle pliers are used by the most expert of fly tyers but they need to be of high quality. Poorly manufactured hackle pliers are the bane of tyers. Nothing is more frustrating than, after much skill and patience, the hackle pliers give up the ghost and let the hackle unwind. When that happens it pays to be out of earshot of the tyer, never mind if he's a priest. Hackle plier jaws must be perfectly parallel with generous finger opening and they must provide good spring strength. Settle for nothing less and buy the best. Another type of hackle tip holder are the push-grip miniature pliers used by radio technicians. When the item is pressed a small hook appears and, when retracted, it collects the hackle. Definitely recommended.

Sharp, fine **scissors** are the good buddies of the fly tyer, the work horse

Perfect pocket water.

of the fly bench. During the process of tying a trout fly scissors will be one of the tools most frequently used. Especially designed scissors are available, with extra large finger holes, and they never leave the tyer's hand from start to finish. Professional fly tyers use them and no doubt they cut down the time needed to tie a fly. But for the home tyer a pair from the chemist will do just as well. Just make sure they are extremely sharp, have fine points that meet exactly, have short blades and long handles with adequate finger holes.

The other tool most frequently used is a razor or any knife with a super fine edge. The former uncased is dangerous and the later clumsy. Far better is a surgical **scalpel** blade. Handles and blades are available at chemists but whereas the blades are cheap the handles are expensive. With a bit of ingenuity an adequate homemade wooden handle is just as good. Mine is still going strong after 30 years or more. Just a word of warning. Take extra care when using this sharp tool. Only too well I recall working at the fly bench when I let the scalpel slip. When I looked down it was quivering, and erect in my thigh.

*When he himself, might his quietus make
With a bare bodkin?*

Certainly, Shakespeare never envisaged his **bodkin** used for such fanciful work as tying trout flies. But used for much less than drama the bodkin is the fly tyer's 'jack-of-all-trades' and useful for a variety of tasks, from picking out varnished hook eyes, to fluffing out thorax dubbing. Stiletto is a more modern term for this handy item but I find 'ye old bodkin rather quainty'. Companion to the bodkin is another useful little tool. I can almost weep when I see some fly tyers using fingers to pick up hooks when a pair of fine pointed **tweezers** will do the job much better. During fly tying sessions I continually stress the usefulness of this little item. As with hackle pliers the points should meet perfectly and should be reasonably robust. My tweezers are 12cm in length which allows for comfortable handling.

Fly tying by daylight is ideal but we are not always so fortunate. Available time is often in the evenings when fly tying can be quite a strain on the eyes. Good lighting is essential. The use of a **spotlight** focused on the vice jaws without causing glare is not a luxury but an investment in your most valuable assets. Many tyers get by without the aid of a **magnifying glass** but for most of us, sooner or later, eyesight becomes a problem. When young a glass may seem superfluous but even young eyes can tie a neater fly through a magnifier. A homemade rig using a small pedestal, clamp, and a normal reading glass is ideal. Unless a back injury prevents it, fly tyers spend a lot of time on their bottoms. Good back support and care in siting the fly bench or table will save much future misery. The ideal bench surface should be roughly 750cm from the floor and the chair cushioned back and below.

I hereby name thee **pigtail**. I created this small but very useful device many years ago adapting a woman's large hatpin to hold the bobbin holder away from the vice. Easy to use, hard to describe. Basically a curly pointed pin attached to a separate post with an elastic band. Heat the end of the pin to red hot and, with fine pliers, twist the point. Make one and bless me, it will be your third hand. Small blocks of **polystyrene** firmly tacked or taped are ideal for holding hooks and sticking in pointed tools when not in use.

I guarantee that one of the first things a new fly tyer makes is some sort of fly tying **cabinet**. They come in all shapes and sizes, some works of art, others little more than adapted cardboard boxes. But they have one thing in common; they make the fly tyer's life a lot easier by keeping materials under control. My cabinet has proved a great asset over the years. It incorporates a four-drawer plastic subcabinet, the type used for screws etc, obtainable from any hardware store. A rear storage space holds capes and other feathers and a polystyrene-coated bar for completed flies. The floor is covered with white vinyl, a perfect background for tying flies. An attached scrap bag is a must.

Author's fly bench.

TYING A TROUT FLY

(All the instructions assume that the reader is right-handed. Sorry about that!)

A hook (#12), a feather (5cm rooster), and some thread is all we need. There is no need to crack your knuckles; we are not going to play a Rachmaninoff concerto. Relax. Make sure the vice is firm and the stem will not twist. Place the **hook bend in the vice.** Tight! Some tyers bury the barb which prevents pricked fingers, others keep it free to prevent damage to the barb. I prefer the latter. I can lose fish easily enough without a damaged barb doing it for me.

Everything needs a start and a finish. A trout fly is a piece of thread tied to a hook, some material added, and the surplus thread removed. To attach the thread to the hook shank pull and stretch some 5cm of thread across the top of the hook shank near the hook eye. Take a turn around the shank but on the next turn **trap the preceding one.** And repeat with a further three or four turns. You can now let the bobbin swing.

Pluck the hackle from the cape and strip away the soft fibres (not all) from the butt. Place the stripped butt **under** the shank and **between** the last thread turn and the eye. Now wrap the hackle stem tightly with four turns towards the eye. Let the bobbin swing.

Take the tip of the hackle in the hackle pliers jaws and with your finger in the loop wind for **three close spaced turns.** Voila! Before your very eyes a collar or ruff is formed. Secure the hackle tip with three turns of thread, cut away the surplus, take another three thread turns behind the eye and, after cutting the thread, the fly is finished. Is this it? I think not! Cutting the thread at this point would be a minor disaster. We have to tie a knot at the end of the operation. It is called the **'half hitch'.** Loop the thread over itself, slip the noose over the hook eye and pull tight. Repeat a couple more times. Another, more sophisticated method is using a **'whip finish'.** It's a little more complicated but in my opinion no better than the H-H.

As the detective said, 'Where is the body?' I deliberately omitted it to emphasise that tying a basic trout fly is easy.

Put your first fly away and place another hook in the vice. Trap the thread but don't tie in the hackle. Just **carry the thread to the hook bend** and prepare to tie in some tails. From the same cape pluck out a large saddle hackle from the side (I could say side saddle but I'd better not) and tear off a little bunch of stiff fibres from mid feather. Keeping the tips level place them on top of the hook bend and take up the thread again. Here's the trick. If you try and tightly wrap the thread around the fibres it will push the fibres around the far side of the hook. But if the fibres are held slightly

Trapping turns.

Hackle tied in.

Hackle wrap.

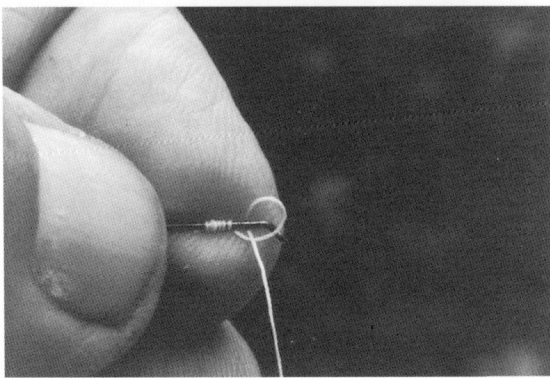

Half hitch.

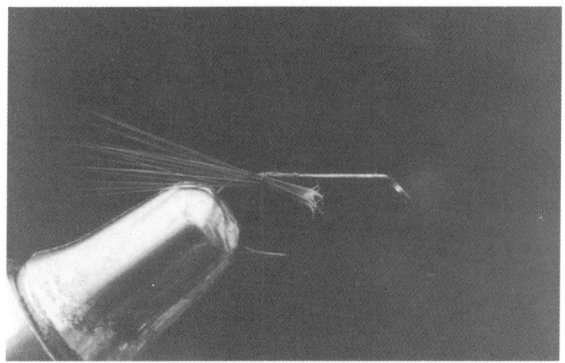

Whip finish.

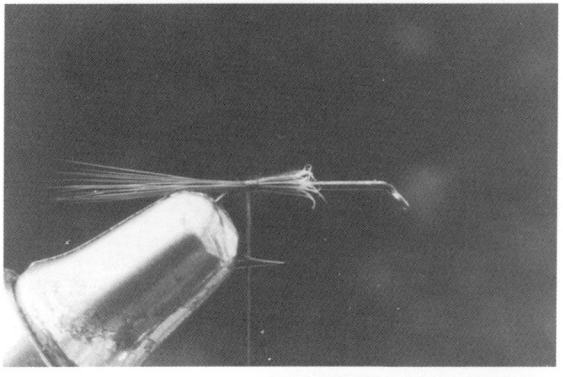

Tie in Tail (a) (b).

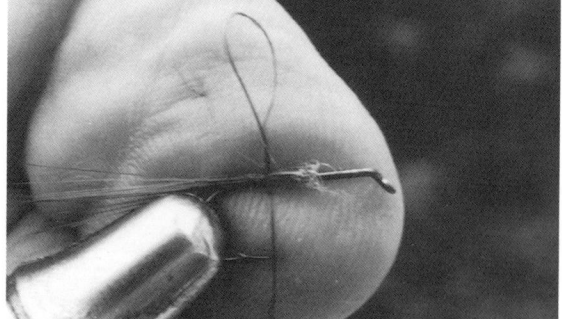

Pinch loop.

to the near side of the hook shank and the thread **wrapped softly around**
before tightening, the tail will be drawn onto the top. Take two more turns,
but before the securing turns, pull the fibre butts towards the eye until the
tail length is correct (tails = hook shank length). Then secure with further
tight turns.

Another method for attaching tails is by using the 'pinch' technique
whereby the tails are held in final position on the shank top, the thread
drawn up between the thumb and forefinger on the near side and held.
The thread is then taken over the shank **leaving a small loop,** the thread
held between hook and finger on the far side. With the thread, hook shank,
and fibres firmly pinched, the **thread loop is drawn downwards** and secured
with further turns. The former method allows the tyer to observe the
operation and manipulate material but in the latter it is hidden.

At long last here comes the body:

This can be formed by using one or more of many different kinds of
material — fur, feather, or silk. Let's start with floss silk and, for good
measure, throw in a rib. This is simply a spiral, usually wire or thread,
wrapped around the finished body. Cut a piece of red floss silk (10cm) and
an equal length of gold thread or wire. (Thread is easier to manipulate).
Tie these in at the hook end and carry the floss silk to a point some two-
thirds towards the eye forming a slim cigar shape as you go. Secure with
three thread turns. **Follow with the gold thread** in three evenly spaced
turns up to the end of floss silk and we are back to the beginning. All we
have to do now is to tie in a hackle and inter the finished fly in the fly box.
And from thence into the mouth of a trout because, despite all the hoopla
and drum rolls, your trout fly will catch just as many trout as a store bought
fly.

But before crowning the fly with the hackle perhaps a little more on
bodies won't go amiss:

Choosing material for a trout fly body is a nice exercise but trying
to imitate exactly a mayfly's translucent tubular abdomen is not only
unnecessary but nigh impossible. Early fly tyers copied the natural insects
as viewed from above and used very opaque materials. Halford, accepted
as a master of the dry fly in the late 19th century, was no exception. Today's
tyers are much more concerned with creating 'impression' than in slavish
imitation. An exception is 'quill' the stripped stem from peacock 'eye' or
a poultry feather. This copies the mayfly's segmented abdomen admirably.
Ostrich herl, with its bulk, is occasionally used but being very absorbent
is little used for floating flies. Wool, once widely used, is now superseded
by the many varieties of synthetic yarns which not only float better but
often incorporate sparkling fibres.

The most popular modern body material is fine hair or 'dubbing'. This
material has the effect of enshrouding the artificial with a 'halo' of refracted

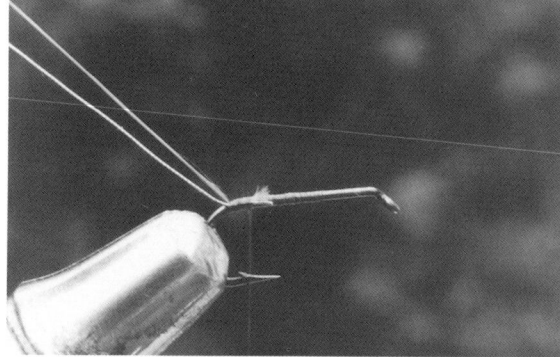

Body materials tied in.

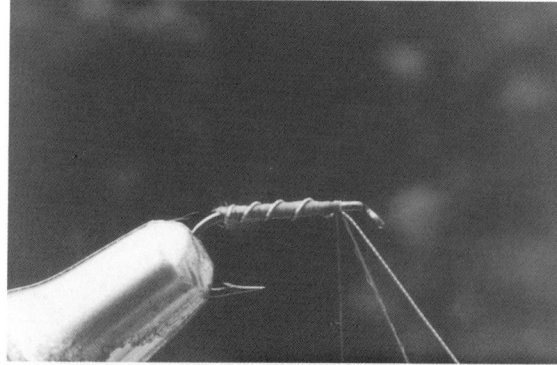

Body completed.

Finished fly.

Red Spinner.

light and even more so when backlit by the sun. Fine plastic strip, which comes in a variety of colours, is popular for some fly patterns and much used in imitations of the stoneflies with their hard, tubular, well-segmented bodies. Imitations of beetles generally use a multi-strand rope of heavily herled peacock 'eye' fibres folded over bodies of some bulky material such as acrylic wool. This material is also used on dry fly patterns. A few patterns call for the use of cork but this is now uncommon.

You are probably impatient to finish off the fly so tie in the hackle as before, form a neat head, and anoint it with a drop of nail varnish. The dry fly of which you are now proud possessor is called Red Spinner.

 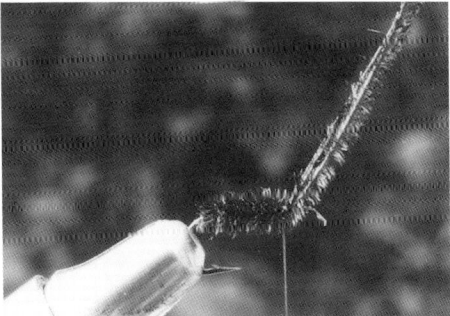

Peacock herl body.

Quill body.

Many other dry fly patterns use fur as the body but it is in imitating nymphs and other subsurface insects that fur comes into its own. In tying fur bodies it is essential to master the technique of **'dubbing'.** There are two methods of preparing this small fur rope ready for wrapping around the shank, the first perhaps the simpler. Using rabbit fur as an example, **cut a clump of back fur** from the skin and from this take a small pinch. Place this on the centre of your palm and gently tease in a circular fashion. With the tying thread in position near the hook bend, and held tight, wax the thread (I use ski wax as beeswax is usually too hard) and **spread a small amount of fur,** thin at the hook bend and thickening away from the hook.

Fur dubbing.

Make a fur rope by twirling the thread and fur clockwise between thumb and forefinger. Wind the tapered rope to the hackle position adding more dubbing if required. With practice this is rarely necessary.

The other method is to create a long loop of thread below the hook bend, **dub one side of the loop with fur** and after pulling the loop tight

Dubbing technique.

with a dubbing twister spin the rope. This traps the fur securely. An advantage of this method is that a space in the upper loop can be left free to add longer guard hairs which when wound protrude from the finer fur forming realistic abdomen filaments. Two tone furs are effective in nymph patterns — abdomen light, thorax dark.

CLIPPED HAIR BODIES

These clipped deer hair bodies are noted for their excellent floating qualities. This is often attributed to the fact that the hairs are hollow. This is not the case. Under a high power microscope endwise (quite a task) the inside of each hair is packed with translucent filaments a matrix with no apparent interstise. It is this pithlike mass inside a water resistant skin that gives the hair such buoyancy and not trapped air. A good example of the clipped hair body is the Irresistible, an American designed fly of which more later.

From a deer hair skin clip a clump of hair and tie in near hook bend. Return thread to near eye and with very secure turns tie in another clump of hair, butts forward, but instead of restricting the hair fibres to the top of the shank, let the tightening turns *ROLL the hair* around the hook shank. Then with thumb nail and forefinger push this collar to the rear of the hook. Repeat the process until a substantial part of the hook is packed with fibres and secure. Now comes the delightful part. With sharp fine scissors trim

off the bulk of the hair leaving whatever shape you desire or as recommended by the pattern.

As a mate to the Red Spinner we can tie a Hare and Copper, one of the most successful nymphs in the fly box. Nymphs work most effectively when fished well down. This means a little lead wire. You can buy it at a tackle shop or, if you can find one, extract the core from an old lead line. Varnish the hook shank and after flattening the end with pliers **wrap the middle third of the hook shank** and break off the remainder. Flatten this end with your thumb nail. Secure the thread behind the eye and cover the lead with tight turns down to the bend. For the tail, tie in a pinch of guard hair, clipped from the back of a hare skin (long fawn fibres), tease out the

Hare & Copper.

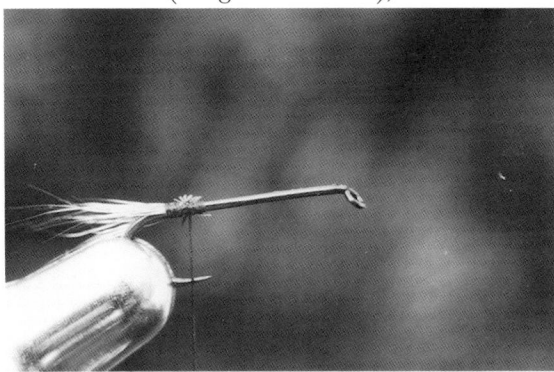

soft underfur and tie the clump in **using the pulled down loop method.**
Tie in a short length of gold thread. Prepare fur for dubbing using light
fawn hare fur and, using either the long loop or the twirl method, form
a tapered fur body covering the lead wraps. The body should be 'clumpy'
at the eye end. Follow with the gold thread (evenly spaced turns) and **tie
in another guard hair clump** just behind the hook eye sloping to the rear.
Form a neat head and afer finishing with three half hitches, cement with
varnish. That's it. A well-tied Hare and Copper which will hold its own
on any trout stream.

These basic ties will provide a good foundation for further fly tying
efforts. Many other patterns require the above basic techniques and as your
tying improves so will the variety of trout flies you produce. A word of
warning here. Keep to standard patterns initially as they are tried and tested.
Some success with them will encourage you to continue with fly tying, after
which, through personal observation of trout stream insects, the urge to
experiment will be inevitable.

Many standard dry fly patterns include paired and matched wings.
This is because wings are always present in the adult form of mayflies, a
common trout stream insect. Then, logically, all artificial dry flies should
have them. But the trout conception of wings may well differ from ours.
After all he views the surface mayfly through a vastly different medium,
especially if the surface is ruffled. We see the fly through reflected light,
the trout through refracted, and possibly only a hazy image. The traditional
materials for dry fly wings are small segments from the grey quill wing of
the mallard duck. These are opaque although probably less so when viewed
from below the surface. Other materials also used include bunched feather
fibres and hair, usually from some deer species. These seen against the light,
are more translucent. Some fly fishers disdain completely the use of wings
and consider a hackle to provide the effect just as well. Right or wrong,
a fly sporting wings does look nice, somewhat like wearing a tie at a wedding.
It is also a skill that, once mastered gives great satisfaction.

Traditional feathers for dry fly wings range from starling and blackbird
to pheasant and mallard wing quills. The latter are most commonly used.
Preparation of the wing slips is very important. The first step in winging
is to **prepare the wing slips** from the quill feather. Wing slips should be
in proportion to the hook size, and be well paired or matched. Using a
mallard quill as an example the softer base fibres should be stripped. Using
a stiletto, **pierce the quill fibres,** choosing a suitable slip width (4mm for
a #12 hook) and by sliding the needle through the fibres separate the slip
taking care not to disturb the fibres. Taking the slip very firmly in thumb
and forefinger **tear it from the quill** with a twisting motion. This helps
to keep the fibres locked whilst tying. Repeat for the other matching slip.

Take the two matched slips (tips splayed outwards) in the thumb and

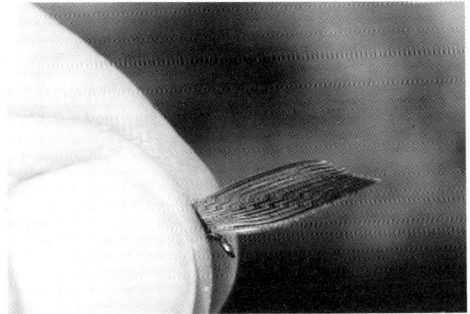

Left: Wing slips prepared.
Right: Wings pinched down.

forefinger with the **tips over the eye** and position them on top of the hook approximately ⅓ back from the eye, keeping the **straight slip edges parallel with the hook shank.** Bring the tying thread up through the fingers, allow a small loop to form above the slip fibres, (similar to tying in tails), pinch this and slip the taut thread down through the far side of the fingers and pinch again. Now for the clever part.

Without hesitation and with plenty of confidence **pull the thread firmly** but not too hard, crushing the fibres in successive layers without twisting them. Do not pause but **take four more turns** trapping the wing base. Throughout the operation it is vital to prevent the slightest movement of the wing slips by keeping up pressure. Only after the last four turns can this be released and even then the finger hold should be relaxed gradually with a gentle upward stroke. At this stage the wings will be upright and if latched together should be gently parted with the stiletto. Take the thread

Left: Wings cocked.
Centre: Wings split.
Right: Figure 8 turns.

forward between the wings, down under the far side of the hook, up the near side, then back rearward through the wing to finish off the figure eight wrap and secure with one turn **behind the cocked wings.** Take the thread once again forward through the wings and take three or four turns directly up to the wing butts, lifting them vertically in the process. Return the thread rearwards through the wings for a final two turns at the rear wing base. The thread can then be taken to the hook bend in preparation for tails and body. With practice the tyer can visualise the slips hidden in the fingers and exactly concertina the fibres.

Feather slips are not the only winging material. Bunched hackle fibres make good wings. A small bunch can be tied in with tips pointing over the eye, raised upwards with a few turns tight into the fibre butts, split and spread with figure of eight wrappings. These wings are more translucent than the opaque wing slips and more than likely, to a trout, resemble a mayfly's wings. 'Antron' (or 'Hi Vis') is a non-absorbent synthetic fibre

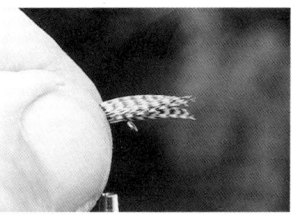

Left: Wing slips prepared and ready for tie.
Centre: Wings pinched down.
Right: Wings upright and split.

suitable for winging dry flies both for duns and spinners. This crinkly sparkling yarn is easy to tie in and is very durable, a wisp of yarn (err on the sparse side), is clipped from the hank and tied in at the thorax area using the up and down loop technique and then raising the yarn to the vertical by butting four or five threads in front. If needed a figure of eight wrap is taken to split the yarn into equal parts. The final stage is to cut the rear excess yarn and trim the wing to the correct height and shape.

Hair wings are relatively simple to tie using hair from fallow or Virginia deer. A pinch of fur is cut or plucked from a skin or pelt and placed with tips forward on the hook shank. Hair is slippery material and should be secured with at least six turns of thread with the forward tip length approximately the length of the hook shank. Keeping the thread tight, the rear excess butts can be cut close to the securing turns. The forward hair is then lifted and secured upright with tight thread turns at the forward hair roots. If a hackle is not used then the hair should be allowed to flair with fibres reaching the horizontal. Flies hackled by this method are known as compara-duns. More of them later.

Wet Fly (wings)

The process of winging wet flies is not quite as precise as for dry flies but it still needs some skill to create sleek neat wings. Any quill wing slips can be used and, as with dry flies, the mallard duck is a good source along with grouse and other game birds. Wing slips of the desired width are torn from the quill but there is no need to use opposite pairs as in dry fly wings. Instead of 'back to back' to create flare they are paired with concave sides together. After completing the fly body, slips are placed between thumb and forefinger, parallel to the hook shank with tips to rear, and with the

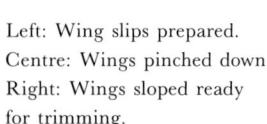

Left: Wing slips prepared.
Centre: Wings pinched down.
Right: Wings sloped ready for trimming.

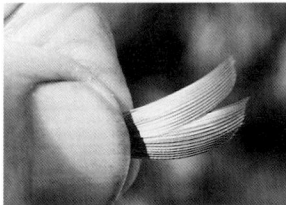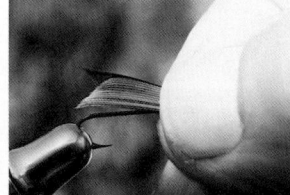

slip butts exposed near the hook eye. One **loose** turn is taken around the butts using the pinch technique to secure the slip then four more increasingly tight turns are wound **towards the eye.** The reason for the loose first turn is to prevent the wings from cocking. Wet fly wings should be streamlined.

Wet Fly (hackle)

Unlike the dry fly hackle, the wet fly hackle is tied in before the tails and body. It does not support the fly but partly masks the body and hook bend, providing a streamlined form as it slips through the water. After the hackle has been prepared by stripping off the 'flue' or down from the feather base the hackle tip is pinched between the thumb and forefinger and the bulk of the fibres stroked towards the butt. The tip is tied in near the eye using the 'under/angle' technique.

After the hackle has been wound with 2½ turns, secured, and the butt clipped off, the fibres are gathered and drawn down and to the rear forming a beard and the thread returned close to the eye. Unless the hackle is wound close to the eye the turns securing them will be too far back on the finished fly leaving an undesirable elongated 'goose' neck. Another method is to strip fibres from the hackle stem and tie them directly under the hook shank behind the eye using the 'pinch' technique. This makes for a less bulkier throat hackle or 'beard' and is better accomplished by temporarily turning the hook upside down in the vice.

Weighted Nymphs

When trout are lying deep it makes good sense to have the nymph sink to near his level. Wrapping the hook shank with lead wire is the usual method or, if not too deep, with copper wire. For standard nymph patterns a few turns of wire are wrapped around the hook and secured prior to dressing the nymph.

Method 1. Smear hook shank with old thickened nail varnish. Do not remove the wire from the spool but flatten the end using suitable pliers. Press this to the shank to the rear of the hook and wind to the thorax area where extra turns are used leaving a slight bulge. Break off the wire and

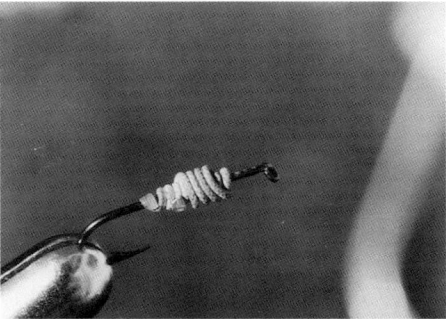

Left: Lead underbody.
Right: Lead underbody with thorax bulge.

smooth point with thumb nail. Attach thread behind eye and take tight wrapping turns over lead to tail. Disregard neatness but ensure good cross wrapping thread turns.

Method 2. This allows for a flattened body, appropriate for stonefly nymph bodies. As before, coat shank with varnish and flatten end of lead wire. Place along the side of the hook shank and wrap to thorax area. Break off and tie in. Repeat the process on the far side. Adjust strips and wrap tightly with thread. Provide another coat of varnish and carry thread to tail area.

Left: Flatform lead underbody nearside. Right: Completed lead flatform body.

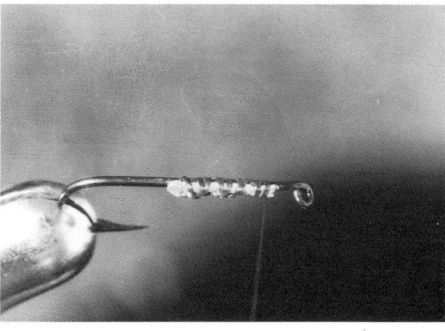
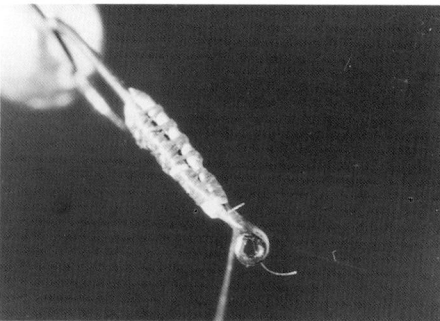

The preceding instructions cover basic fly tying. More advanced techniques, such as in tying parachute, hair body, compara-duns, emergers and other styles are fully described in later chapters.

Author at his fly bench.

8

Trout Stream Insects

Many fly fishers spend their angling careers blissfully unaware of the hidden world of aquatic insects and yet continue to catch trout on a great variety of artificial flies. As long as this or that fly can catch trout, why wonder what they are supposed to copy, if anything? Some fly fishers may harbour a niggling doubt that knowing a bit more of the trout's diet might help them catch more fish. Others are curious but are put off by scientific jargon. The British have no such problems giving common names to natural trout food insects. But there *is* a particularly good reason to study aquatic insects, even briefly, and it is this: Because of increased fishing pressure, our trout are becoming ever more selective which in turn makes choosing your artificial even more important.

Aquatic insects are vital to trouts' survival. Terrestrial insects are less important but they play a significant role. Other trout food items such as bullies, crayfish and worms are catch crops and whitebait is seasonal only. These, together with crustaceans and small bait fish, provide food for esturine trout. The river bed provides a home for vast numbers of insects of many shapes and sizes. They may be large or small, soft or hard, immature or adult, scarce or plentiful, swim freely or be reclusive, crawl or swim, but each will be either prey or predator. At the near end of this lengthy food chain is that sleek but savage beauty, the trout. Sharing the final niche with shags and eels is the angler, and with his tiny flies the most deadly predator of all.

Following is a list of trout food insects omitting uncommon food items and, generally, in order of importance.

CADDIS FLIES

This group is commonly found in trout stomach autopsies. Of the many varieties two are significant from the angler's viewpoint. The **cased** caddis and the **free** swimmers both in the juvenile and adult stages.

Cased Caddis

Easily recognised by the horn-shaped tubular capsule, cased caddis provide the trout with a reliable source of food. Inside the case the small grub is usually safe from predators but the hard, painstakingly constructed shell is no protection from the trout's acidic stomach juices. These tiny insects are amazing builders. Starting with a fine, extruded sticky thread, they spin a cocoon, while at the same time selecting either fine gravels, grit, or tiny pebbles and attaching the material to the outer case, extending it as they grow. Depending on the species, a variety of stream bed material is used, from coal specks to mica fragments. The chance of finding one covered in gold, while an intriguing prospect is surely in the realms of fantasy. But even more bizarre is the species that covers its case with leaf fragments, each tiny piece being cut with equally tiny mandibles to form an exquisite and colourful tubular mosaic jigsaw.

Free Swimming Caddis

Unlike the cased variety these soft-bodied larva-like juvenile caddis are relatively active. The woolly caddis, a dingy grey larva, builds fine mesh nets between tiny rock or debris crevices. Periodically they emerge, like some aquatic spider, to scrape minute food particles from the trap. It was in *Trout Stream Insects of New Zealand* that I gave this larva the common name of 'woolly caddis'. This because of the bushy abdominal gills. It is one of the few aquatic insects that can survive in poorly oxygenated water such as that found in backwaters. Trout are not averse to grubbing among the cobbles and stones for these plump larva.

Other caddis genus of importance to the angler are the smooth, green decorated caddis larva which appear in various hues from ochre to lime or grass green. Most of these larvae sport intricate dorsal patterns but, with their formidable mandibles or pincers, they all prey on smaller insects. They have a very distinct abdominal segmentation, a characteristic which the fly tyer can imitate to advantage.

Adult Caddis

The winged stage of both cased and soft caddis are easily distinguished from mayflies both by wing shape and flight character. The latter sport upright gossamer wings whereas the fine hair covered wings of the caddis slope over the abdomen. Mayflies have a drifting determined flight pattern — caddis flutter all over the place. It is a wonder they ever get there. While trout feed mainly on juvenile caddis, egglaying adults are eagerly sought after especially during late evening hatches.

One of the best hatches of caddis I have ever seen occurred up the bush section of a boulder-strewn West Coast river. My tally for the day had been only two fish, an eight pounder and another slightly larger. I'm allowed to skite just a little bit! Both were taken on Darter nymphs and the last thing I expected was a caddis hatch. Around nine o'clock I was ready to climb up the steep bank to the caravan when the first caddis fluttered on my cheek. Within a minute hundreds of them swarmed over my hat, jacket and hands. Upstream, the mirrored surface of the pool seemed to be boiling as trout feasted on the tiny insects. I was so excited I could hardly cast but after dozens of casts I failed to connect. Before darkness sent me packing I had only one fish take a dry fly pattern. I did learn something though. Don't count your chickens etc etc and, always carry a good caddis pattern!

MAYFLIES

Watching an adult mayfly, with its delicate gauzy wings fluttering in a gentle breeze, it is difficult to picture it in its relatively grotesque aquatic nymph or juvenile stage. Mayflies, with lifespans varying from around 12 to 24 months or more, spend little of that time in our atmosphere, living nearly their entire lives underwater. As with caddis some are prey and some predators but all proceed through many nymphal stages or instars. Unlike we humans, whose skeletons grow as we age, mayfly nymphs are not so lucky. Their external body case is their skeleton and as it cannot grow it is split and discarded at various periods until the mature nymph is ready to surface, change form, and take to the air as a winged adult. Mayflies are unique in the insect world, having an intermediate stage known, in angling parlance as 'duns', before finally emerging as 'spinners', the final, reproductive stage. Although New Zealand hosts around 14 known species of mayfly only about half a dozen are of significance to the fly fisher.

Small Brown Nymphs

The most common (and here we must give the true name) are the two *Deleatidium* species, *vernale* and *myzobranchia*. Also of importance are species of the *Zephlebia* genus which are small but vary in appearance. Unless of an entomological bent these are not so important except perhaps for a North

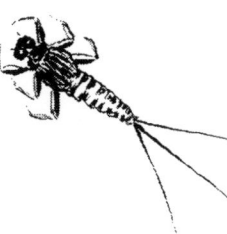

Darter nymph. (Nesameletus sp)

Adult Darter. (dun)

Kakahi Queen nymph.

Kakahi Queen. (dun)

Small brown nymph.
(Deleatidium sp)

Small brown dun.
(Deleatidium sp)

Small brown spinner.
(Deleatidium sp)

Gravel Cased Caddis.

Green Caddis.

Darters *(Nesameletus sp.)*

Nymph.
Very common in nymphal form the darters are appropriately named and easily recognised by their streamlined torpedo shapes and dark-banded, fringed tails. The head is helmet-like with large dark eyes, the legs banded, the abdomen multi-coloured and well patterned. They have very active paddle-shaped gills. Needing well oxygenated water they mainly inhabit the faster stretches of a river and, unless swimming, which they do extremely rapidly, they are exceptionally well camouflaged. The duns, larger than Dad's Favourite or Early Browns, are very pretty insects with heavily mottled wings (with the leading edge yellow) and colourful abdomens which end in a dull white segment. They have ochre-coloured legs, each with a prominent black band and three tails, the centre one rudimentary and the outers, fawn with dark bandings. The spinner, apart from the paler abdomen and clearer wings, is very similar to the dun. Both carry the pale eighth segment and, although less obvious on the spinner, sport the lovely yellow wing stripe.

My pattern, the Grey Darter is a tried and tested imitation of *Nesameletus* and a variation, the Blue Darter, is even more effective.

Dun. Kakahi Queen Hook, size 12-14-16: This is a very pretty mayfly with a patterned olive body and well veined wings with a distinct yellow leading edge. Legs and tails are straw coloured.

Spinner. Twilight Beauty. Hook size 12-14-16. Similar to the dun but with brilliant clear wings which still retain the yellow wing stripe.

Kakahi Queen Nymph *(Colorburiscus humeralis)*
The only specie of its genus, *Colorburiscus humeralis* (from now on Kakahi Queen Nymph, hook size 12-14) is a plump, dark-brown, hunchback nymph, named by Charles Nott of Havelock North who supplied the entomological work for the classic *Trout with Nymph* by Tony Orman. Before describing this nymph and adult it may be helpful to provide some background. The dun had long been imitated by the late Basil Humphries of Kakahi who gave it the delightful name. Nott, appropriately, connected the nymph to the common name of the adult. Only its mother could call this nymph handsome. It has been described as an aquatic Quasimodo, the hunched thorax, clawlike legs, and rows of spiny appendages above the abdomen giving it a rather menacing appearance. But, like many fearsome looking creatures, it is a vegetarian with gills often draped with tiny items of stream debris. It is a poor swimmer and having immobile gills needs a constant and plentiful supply of oxygen. It cannot survive in still waters. The predominant colours are deep sepia and chocolate brown with three fawn tails, the centre one rudimentary.

The dun and spinner are very similar to those of the Darters both

sporting yellow wing edges but there are two obvious differences. They are distinctly larger and the Kakahi Queen (Hook size 10-12-14) dun displays two glorious bright yellow blotches at the base of each forewing. Less so, but still obvious on the spinner. On rivers such as the Clutha there is sometimes a spectacular hatch of these large spinners. I well remember an evening fishing with my son, then working as a wildlife fishery cadet at Wanaka hatchery. I had almost given up hope of any fish rising even after his urging to hang on a while. Just on dusk came a tiny ring way out in the current. It looked like a tiddler rise. Another followed, then another until the river surface boiled. What a night that was. Before it turned really dark we landed eight fish between us totalling over 20kg. An autopsy on two of the fish kept for smoking showed the fish gorged on Kakahi Queens. When these spinners are about, one of the best imitations is the Twilight Beauty (Hook size 10-12).

STONEFLIES
Brown Black and Green

Generally regarded as back country river nymphs these little carnivora are also usefully imitated and fished on lesser down country waters especially in the upper reaches where rapids and riffles are frequent. Stonefly nymph species vary in size from 10mm to 30mm and, apart from a unique green species, come in various shades of brown or dark umber. Similar to the mayflies they progress from nymph to adult but without the mayfly's intermediate stage. Easily recognised by their outrigger legs, prominent wingpads and well segmented bodies, stonefly nymphs prey on anything they can get their stout mandibles into from sandfly larva to tiny caddis and mayfly nymphs. All stonefly species have only two tails.

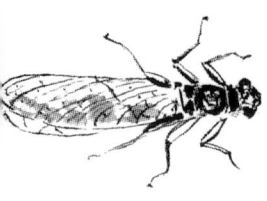

The green stonefly nymph hardly needs description with its colourful abdomen, green on the dorsal and salmon pink on the ventral. Even the action of stonefly nymphs gives the game away — they waddle. After performing many trout autopsies I can count on one hand the number of *adult* stoneflies found in trout stomachs. Obviously the female must return to the river for egg-laying but it may well be that, unlike the mayfly which expires on the water, the stonefly adults depart for another cemetery.

Now I hang my colours on the mast. I suggest that the fly fisher who is able to identify, even roughly, the insect on which the trout is feeding and then select a fly that closely imitates it, will not only catch more trout but increase the pleasure ten-fold.

9

Other Trout Stream Goodies

DOBSONFLY *(Chauliodes diversus.)*
The ugly duckling of trout stream insects and the scourge of other river bed fauna, the dobsonfly or creeper, as it is commonly known, is the largest of our stream bed insect inhabitants. When mature, the larva of the dobsonfly is a formidable insect. The larva may be ugly but beauty, as they say, is in the eye of the beholder. No doubt during the mating season *adults* view each other with tender eyes but they rarely provide food for the trout and are not worth imitating. Not so the larvae which, when uplifted during spring and early summer floods and washed downstream, are a prime source of trout food. The dingy grey, black-headed larva, with fingerlike lateral gills and strong mandibles (miniature pincers), spend a long, active hunting period before migrating to the beach. After finding a suitable riverside stone they settle under it in moist or wet sand and create a small grotto to pupate. Sometime afterwards, usually at dusk, the adult emerges from the pupa and takes to the air in clumsy flight. The female, with a 70mm wing span, when encountered at twilight, can give the angler a fair old fright. Used as bait they are deadly as trout catchers but the Woolly Caddis (described later) makes a good imitation.

Dobsonfly larva.

Water Boatmen.

Passion Fruit Hopper.

Daddy Long Legs larva.

Damselflies. From trout's stomach.

Cicada.

Willow Grub in gall.

Woolly Caddis.

Green Caddis and pupa in
capsule.

Other Trout Stream Goodies

Green Stonefly. (also ventral view of Kakahi Queen nymph)

Peter O'Rourke nets a fine West Coast brown.

BEETLES *(Coleoptera.)*

Seasonal and non-aquatic, these insects provide the trout with a substantial part of his annual fare. December through to February are peak months when large numbers of beetles find their way onto the water. The fly fisher is mainly concerned with two types, the brown and the green. The former, commonly known as the grass grub, the latter known as the scarab or green beetle. The main characteristic of both is the hard overshell which is really one pair of the insect's wings. Trout are particularly attracted to beetles. Once on the water they are captive and provide the trout with an easy meal and, as any fly fisher can testify, during those periods the trout are easily fooled. Beetles can arrive in great numbers. I recall staying at a fishing hut during an explosion of brown beetles. A window had been left open and I had to use a broom and shovel to sweep them out.

Beetles appear in trout stomach autopsies throughout the fishing season but especially during the emergence period of grass grubs. Anglers who are fortunate to be on the river when these plump terrestrials settle or are blown onto trout waters can be either blessed or frustrated. Trout gorging on beetles, with hundreds of naturals to choose, may well ignore the artificial but sometimes perseverance is rewarded.

Brown Beetle *(C. zealandica.)*

These are a pest and few appreciate them. The exceptions are the trout and the angler. After a long period feasting on the farmer's paddocks they emerge in great numbers and, inevitably, a good percentage land on the waterside. Here the trout are ready customers. Brown, hard-shelled with little features they are one of the easiest trout food insects to imitate.

Green beetle *(P. festiva.)*

Commonly known as manuka beetles these chunky insects with the brilliant iridescent green wing cases are welcomed by the fly fisher. Whether solitary or in summertime masses trout very rarely fail to take them and they warrant a place in every fly box. Also known as scarab beetles they were used as decorations by the ancient Egyptians, examples of which can be found in natural history museums.

On the subject of beetles I remember one of those special occasions when the trout literally went mad for these hard-winged insects. The Mangles, a few kilometres north of Murchison, is a curious river, the lower eight kilometres running through a blackberry and scrub-covered gorge, the upper flowing through open paddocks and low hills. Fishing had been just so-so and Mark, my American friend, was beginning to have doubts. Around mid morning the breeze strengthened to a strong wind and we witnessed what appeared to be a miracle. Upstream in what seemed to be a fishless pool, the surface literally boiled, trout rising everywhere. With

Collecting nymphs.

squadrons of brown beetles drifting past our legs we tied on imitations and proceeded to have a bonanza. But above calls of 'Nice one'!, 'Wowee!' and the whirr of reels came shouts for help. We paused, listened and then dashed upstream to find his wife in trouble. She had decided to cross the river, reached the point of no return and was standing petrified in the middle of a fast rip. Mark dashed in up to the waist and grabbed her arm while I followed and, with her feet floating, we pulled her out. After that experience it was back to the van and dry clothes then dashing back to the river and those mad trout. The beetle flight was over but we did have a consolation prize picking up a couple of fish mopping up dead beetles in tiny back eddies. A very exciting hour.

WATER BOATMEN *(Sigara sp.)*

All anglers are familiar with these small oarsmen that frequent silty stream edges and backwater shallows. During high summer they provide the trout with a substantial part of his diet. A striking feature of these insects is the disproportionately long fore and median legs, which, with fine hairlike fringes, give excellent propulsion. Boatmen are air breathers and must rise to the surface at frequent intervals and in doing so collect small air bubbles used as a reservoir. This is a most excellent feature to imitate by including some form of silver material in an artificial. On occasions I have seen cruising backwater trout actually ploughing the silt then circling back through the cloudy water to pick off disturbed boatmen. Similar to beetles they have hard protective overwings with softer flight wings underneath.

MIDGE PUPA *(Chironous sp.)*

One of the smallest of trout stream insects, they make up for it in numbers. Millions of these tiny buzzers are hatched throughout the season emerging at dusk. The larva of the midge is commonly known as the bloodworm, a tiny worm-like creature that inhabits the silt and mud of the stream and lake bed. The larva is blood red, owing to the presence of haemoglobin, a pigment contained in human blood and other larger animals. At this stage, during floods, when it is wrenched from its secure habitat it becomes a valuable trout food. On emerging as pupa, it rises to the surface and transforms into the adult midge. It is during this pupal stage that trout show avid interest and await the final feast of the day.

The 'mad hatch' that occurs frequently during evening falls of mayfly spinners can also be linked to midge pupa hatches. At dusk, pupa rise in countless numbers popping into the air like polaris missiles. While suspended in the surface film these olive green or claret-coloured pupa with their pale, tufted antenna present an easy target. The water can literally 'boil' and what seems like a promised bonanza to the fly fisher can become a frustrating experience as the trout have thousands of naturals to choose from. What price one artificial among this lot? The adult rarely features in trout fly patterns probably becuse it is so tiny, with the smallest trout fly at least a hundred times larger than the natural. Pupa imitations, however, can be very deadly.

PASSION VINE HOPPER *(S. australis.)*

When these small delta-winged bugs are in season many are blown onto the rivers and streams where the trout will feed on them almost exclusively. Adult hoppers are easily recognised by the oddly shaped speckled wings and their ability to disappear with an incredibly quick hop. They are also easy to imitate. Being light and immobile on the water the hoppers soon congregate into sudsy feedlines making them sitting targets for trout. P.V. Hoppers are found in the Marlborough and Nelson regions and also in parts of the North Island.

DADDY LONG LEGS *(Tipulidae sp.)*

The adult, a true fly (having only two wings) is common both in the home and in the field but it is the larva of this insect that puts it in the category of trout food. Easily mistaken for an oversize maggot the larva has few outward characteristics. They are often found in trouts' stomachs especially after the stream bed has been stirred with floods. Unless a trout has been stationed below a dead sheep, any maggot-like grub in the stomach will probably be a cranefly larva. They are easy to imitate with a very common item described in a later fly tying chapter.

HAWTHORN FLY *(P. nigrostigma.)*

A true fly and its name, although not exclusively, gives its habitat. Somewhat similar in appearance to the common housefly it can be recognised by the exceptionally long drooping legs and substantial probiscus. Another feature is the prominent black spot on the wings. As with all true fly adults what would normally be a second pair of wings have evolved as tiny halters or balancers, like tiny dumb-bells. While not main trout food insects they are often found during stomach autopsies. For the enterprising fly tyer the hawthorn fly presents a nice challenge.

DRAGONFLY *(Odanata.)*

Almost a rags to riches story, dragonfly nymphs, with their fat ungainly bodies and bulging eyes, are not one of nature's prettiest creatures. They are, however, unique in having a grotesque chin or 'mask', normally folded under the abdomen, but which can be rapidly shot forward to secure other insects or tiny fish. Talented fly tyers have produced extremely realistic imitations, taking fly tying to an art form. The adult, with shimmering gossamer wings, is often seen darting and hovering over the stream in search of prey. On slow sections of the river or on ponds or backwaters, cruising trout can be even quicker and with a sudden leap catch these slender aerialists. Again, with the imitation, the fly tyer becomes the artist.

DAMSELFLY *(Zygoptera.)*

The adults of these pretty insects can be seen throughout the summer months flitting around the stream edge or lake margins. They can be recognised by their iridescent metallic blue/purple wings and disproportionate body length. The adults often fall prey to the trout while hovering over the still waters of ponds and lake margins but it is the damselfly nymph that really fattens him up. Unlike the dragonfly nymphs, which are squat and inhabit dense weed beds, those of the damselfly have prominent eyes, long legs, three feathered gills and are slim and relatively mobile. The adults with their large transparent wings perform amazing acrobatic feats securing smaller insects in flight. One species will get no kudos from the lady anglers. When ready to reproduce the male damselfly seizes the female by the scruff of the neck and dunks her into the water for the egg-laying ceremony. He then hauls her back up for another bit of fun.

CICADA *(Hemiptera.)*

A terrestrial, the *nymph* of the cicada is of no interest but the clatter of *adult* cicadas during summer is music to the angler's ears. With glassy wings held tentlike over the abdomen, the hard-shelled thorax, and head parts with prominent eyes, this glittering bug is a favourite food of trout. A large insect which must be the equivalent of a thousand mayfly nymphs or adults, it is no wonder a very large trout will sometimes rise metres from a deep hole to take a juicy cicada. When tying cicada imitations there is no limit to the fly tyer's imagination.

WILLOW GRUB *(P. proxima.)*

These small larvae can be a fly fisher's curse. Once trout lock on to willow grubs they become very selective and it is the devil's own job to catch them. The sawfly adult injects an egg into a willow leaf where it pupates inside galls, recognisable as small rosy red blisters. After emerging as tiny, yellow, black-headed grubs, they lower themselves on a silk-like thread to the ground or, by misadventure, onto the stream surface. Once on station under the willow branches a trout will feed almost exclusively on the floating larvae and it is not uncommon, as I well know, to spend embarrassingly long periods without success. The adult sawfly is of little interest to the fly fisher.

Usually, the time of season, or habitat is a good indication of which type of fly to use but there are occasions when all our theories get turned upside down. There is no doubt about it. The longer we fish the more we realise how much we have yet to learn about the trout. One day he appears

to be silly and the next as crafty as can be. A few weeks ago, fishing with my friend Len Prentice, we had a remarkable experience while fishing a Southland river. It was my turn to try for a lovely brown cruising around a willow-lined backwater while Len perched on a nearby branch for a grandstand view. Try as I might I couldn't interest that fish and, after an embarrasing period, I began to despair.

When Len suggested using a lure I thought he was joking but then I decided I couldn't do any worse and tied on a long, shanked damselfly nymph, herl body and a red throat hackle. I fully expected the fish to either ignore it or bolt for the willow roots but as I jerked it alongside the weedy edge a bow wave appeared and Len gave a great shout, 'He's after it, Norm!' And the next minute the trout hit the lure. Smash is more like it. I didn't have to strike. He hooked himself and fled across the pond. Another fish appeared and followed his every twist and turn until, in the now soupy water, the net enfolded him. 'Nice fish,' said Len, as he watched his wobbly journey home. 'Now,' I thought, 'I have the answer to these shy backwater trout.' Since that day I have never pulled off the same trick which goes to prove — the longer we fish the less we really know.

After these chapters on trout stream insects I hope that during your next fishing trip, if you come across a fellow angler foraging for insects or shaking a willow branch, you won't give him a wide berth. He may not be as crazy as you think.

Vern Williams with a ten foot rod and a four pound trout.

10

Popular Trout Flies

I would be a rich man if I had a dollar for every trout pattern that I have tied but there would be a poor reward for every 'super fly'. Trout catchers supreme are as rare as dogs in a butcher's shop. More common are 'reliable' popular trout flies as distinguished from custom-built creations. By reliable I mean artificials which catch trout under most conditions, providing the skill of the angler matches the attributes of the fly. The choice is tremendous and it is no wonder the fly fisher's box is full to the brim with more patterns than he can possibly use. Only experience will trim them down to favourites that have stood the test of time but some anglers are relatively new to fly fishing and have yet to start pruning. These newcomers I envy. How wonderful it is to be at the threshold of the sport. Perhaps opening my own fly box would be helpful.

My armoury of flies is by no means infallible. If that were the case I could retire tomorrow and become the most unpopular angler in the business. The charm of fly fishing lies in the unknown and to diminish or remove that aspect would defeat the purpose. Nevertheless the following trout catchers can be used with confidence and that is very important. Time spent constantly changing flies is time wasted. The old saying that 'it's the flee on the water that catches fish' is as true today as it was in Walton's day.

93

A SELECTION OF TROUT FLIES

1. For convenience, all materials for patterns are listed in the sequence they are applied to the hook.
2. Hook is assumed to be in the vice prior to the method description and the thread affixed.
3. Most nymphs are weighted with lead wire.
4. All fly heads are assumed to be cemented after fly is tied.

DRY FLIES

Greenwell's Glory

Although originally intended and used as a wet fly, since its first appearance in England in 1854 it has become a very successful floater. It was probably one of the first traditional trout fly patterns to be used in New Zealand and is a fair imitation of some of the *Deleatidium* species duns.

Hook	12/14/16		
Thread	Olive	Wings	Mallard slips
Tails	Furnace		
Rib	Fine gold thread		
Body	Primose (waxed) or olive floss		
Hackle	Furnace		

This dressing varies from the original which has a Coch-y-bondhu hackle. This has a black centre, red furnace fringe and black outer tips. They are very rare and unnecessary. A dark red or dark ginger furnace is just as good. For some reason the Greenwell fishes extra well when the river levels drop after a good fresh. It is also a very good prospecting fly.

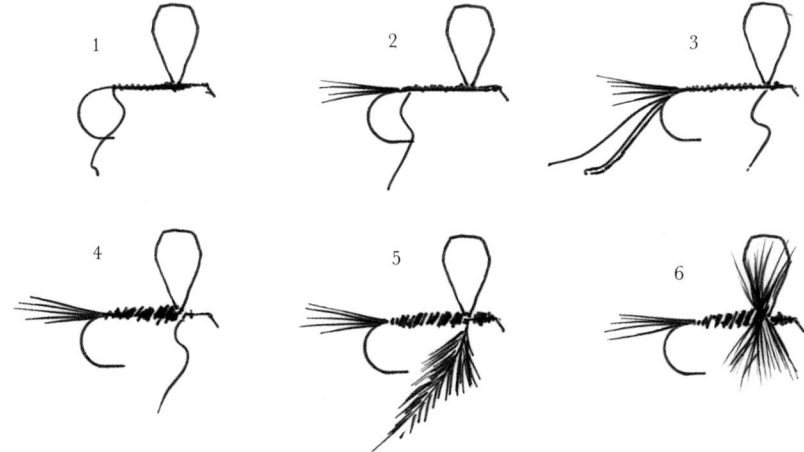

1. Wings tied in. 2. Tails tied in. 3. Body materials tied in. 4. Body formed.
5. Hackle tied in. 6. Hackle wound.

Coch-y-bondhu

I can thoroughly recommend it as being one of the best 'all round' flies. To the trout it may resemble a beetle or it may be that it just looks 'buggy'. Whatever, it is a very good searcher and can stand at least two hackles. I use three on a #12.

Hook	10 to 18
Thread	Black
Tag	Flat gold tinsel
Body	Bronze or green peacock herl
Hackles	Furnace or dark ginger/black mix

This is a 'plump' fly and if tied correctly, in shape, should look like a cross between a bumble bee and a grass grub beetle. It needs to float very proud so pack as much hackle onto the fly as size permits. On larger sizes I tie in an underbody of yellow or red floss silk, which, after the fly has been used a little, tends to show through the herl. The peacock herl strands should be taken from below the 'eye'. The gold tag is one of my additions. This gives that little 'glint' when viewed from below and which I think makes all the difference. For good support I also add a *short* stubby tail.

1. Tail tied in. 2. Body materials tied in. 3. Tag formed. 4. Herl wound. 5. Hackle tied in. 6. Hackle wound.

Red Tipped Governor

This trout catcher can be described as a Coch-y-bondhu with wings. Another searcher, it has the advantage of being easily seen in rough water. Originally an old English fly it gained a red backside on arriving in New Zealand hence the tag. The original pattern includes a hen pheasant wing but these are so fragile I omit them.

Hook	10 to 16
Thread	Black
Tag	Red floss silk
Body	Bronze peacock herl
Hackle	Dark ginger

Whenever I encounter strange water especially if it is fast and stony, I invariably tie on a Governor. On a visit to Britain I used it to good effect on the River Wharfe in Yorkshire. This river is very similar to the Motueka River, quiet flowing with rock outcrops and even if the trout are smaller the fly hatches are superior. Unlike most New Zealand rivers with uncultivated banks and difficult access, many English rivers are bordered with National walking paths and it is not unusual to have a small audience while you fish.

Kakahi Queen

Little needs to be said about this fly as it has been well covered elsewhere except to add that it is a very old friend of mine and has given me some wonderful days on the river.

Hook	10 to 16
Thread	Brown
Wings	Mallard
Tail	Ginger
Body	Striped peacock quill
Hackle	Dark ginger

The only tricky part of tying this fly is in the wings which have a narrow strip of yellow mallard (dyed) near the forward edge of the wings. This is best accomplished by tying them in after the main wings have been set and using the natural curve of the mallard speckled breast feather to keep from splaying outwards. Selecting a well-striped quill makes for a better looking fly but the trout may not be so choosy.

Dad's Favourite

Probably one of the best known dry flies in New Zealand this is also home grown. To some unsung designer we owe a word of thanks. Over the past 60 years or so it has been responsible for the demise of countless trout and no doubt because of its likeness to *D. vernale,* one of our most common duns.

Hook	12 to 18
Thread	Brown
Wings	Mallard (dark)
Tail	Dark ginger
Body	Striped peacock quill
Hackle	Dark brown

It is well liked in Southland's downcountry rivers where the natural mayfly is abundant. The Mataura River is a dry fly angler's dream, especially in the upper reaches around Nokomai. The pools are frequent, usually slow flowing with few rapids, ideal habitat for *Deleatidium* nymphs.

I recall a day on the upper reaches when the fishing had been slow. With only one fish to show for my morning's efforts I was admiring the antics of some fantails when Jean touched my shoulder. 'There's some mayflies,' she said, pointing upstream. Sailing downstream were a few 'vernale' duns but even before I had switched to a Dad's Favourite the air was full of them. Looking upstream against the sun they appeared to be a shower of glittering confetti but my attention was quickly drawn to the river. Trout were rising everywhere, gulping down the helpless victims. The hatch lasted over a half hour but in that time I had landed (and returned) eight good fish. Which goes to show you never know when a 'red letter' day is around the corner.

Twilight Beauty

It may seem strange that, although the wings of the spinner *C. humeralis,* alias Kakahi Queen are brilliantly clear, the wings of the artificial 'Beauty' are dark, near black. The designer, Basil Humphries, knew a thing or two. He realised that when viewed from below in the late evening the trout would see them mostly as a silhouette. Whether on downcountry streams or wilderness rivers the 'Twilight' is a most dependable trout fly.

Hook	10 to 16
Thread	Black
Wings	Mallard (black or dark umber)
Tail	Dark brown
Body	Black floss silk or thread
Hackle	Chocolate brown (dark)

Apart from the wings, which should be tied slightly forward, a distinctive character of this fly is the pronounced thorax. I prefer well-waxed, thick thread for the body as the floss is fragile and after a few good trout have had their way with it it becomes frayed. The thread body is robust and three or four turns behind the wings builds a nice thorax. The waxed thread also helps it to float better.

Coachman

A workhorse of a trout fly, and similar to the Governor, the Coachman is my choice for a good searcher pattern. With a thick herl body and white wings it can be seen under rough water conditions and in poor light. There is an old North country saying, 'If tha' doon't nay wat ta' do, send furt coochman'.

Hook	10 to 16
Thread	Black
Wings	Mallard or goose quill (white)
Body	Bronze peacock herl
Hackle	Red/brown

Trouts' teeth can make a mess of peacock herl and one way to overcome this is to make a twist using a short length of the tying thread. Depending on the hook size three or four herl strands should be used. If green herl is exposed to the sun for a few weeks the colour changes to bronze/purple.

WET FLIES

Grouse and Purple (Soft hackle)

My introduction to soft hackle flies was back in 1947. I was lucky enough to have permission to fish a small Yorkshire stream or 'beck' as they were known. The trout were small, averaging three to the pound and a trout over a pound was quite an event. Harry Clegg had fished it for years and I learned a lot from him. His technique was to fish a wet fly, *upstream*, to rising fish and *stroke* the fly across or near the rise. He was using what is known today as the induced take method of nymph fishing. His flies were nearly all soft hackled such as Partridge and Orange or Waterhen Bloa both of which are just as deadly on our New Zealand trout.

Hook	12/14
Thread	Black
Tail	Black cock
Body	Purple floss
Rib	Gold thread
Hackle	Grouse (back or covert feathers)

For some inexplicable reason trout seem attracted to the colour purple. I know of no trout stream insect that sports that colour so it may show that trout may not see colours as we do. Whatever, I can recommend the Purple Grouse as it is commonly known, a great fly when used as a dropper and with a Dark Red Spinner as tail fly.

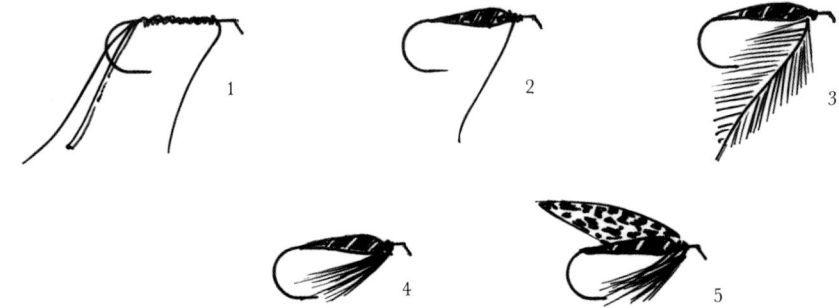

1. Body materials tied in. 2. Body formed. 3. Hackle tied in. 4. Hackle wound and shaped. 5. Wings tied in.

Waterhen Bloa

Soft hackle flies come in a large range of patterns and styles. An effective pattern, the Waterhen Bloa copies well the emerging nymph of *Deleatidium sp.* Any soft grey hackle will substitute. A dubbed body using grey rabbit mixed with Hi Vis or Sparkle Yarn makes a very attractive fly. During evening hatches, when duns are emerging, many of which cannot penetrate the surface, sliding the fly across the trout's vision often results in a 'take'.

Hook	12 to 16
Thread	Black
Body	Rabbit underfur
Hackle	Waterhen (under coverts)

Like many other predators trout seem attracted to what is obviously a 'live' target so perhaps it is the mobility of the hackle fibres that also make this wet fly so successful. The soft underwing feathers of quail or teal can be substituted for the waterhen but both should be spoon-shaped and when tied just mask the hook point.

Partridge and Orange

Fished down and across chuckling rapids of small streams, with a Dark Red Spinner for company, the results can be surprising. The two must present a very cheeky pair as they bobble here and there. I love fishing these intimate waters where every rock and crevice may hold a trout and fished up or down they fare just as well.

Hook	12 to 16
Thread	Orange
Body	Rich orange
Rib	Gold thread
Hackle	Brown partridge

The hackle feather is taken from the back of the English partridge and not the French bird. A good substitute for partridge is a speckled bantam. If you are fortunate to come across one of these small birds and can induce the owner to part with it then, after the appropriate rites, and after removing the wings, skin the whole bird. It will provide you with many years' supply of both winging and hackling feathers suitable for many other patterns. A speckled hen skin is a most prized possession.

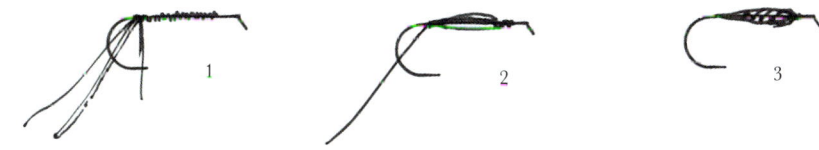

1. Body materials in. 2. Silk body formed. 3. Rib wound. 4. Hackle tied in. 5. Hackle wound.

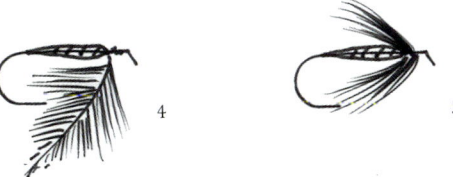

Dark Red Spinner

Although this fly, when tied as a wet, bears no resemblance to a spinner it does catch trout and should be included in the fly box. Highly recommended.

Hook	12 to 16
Thread	Brown
Tail	Brown hen
Body	Claret or crimson floss
Rib	Gold thread
Hackle	Red/brown
Wings	Bronze mallard (breast) tied well sloped.

In winging, a mallard slip (10mm width for a #12 hook) is folded once, employing the natural curve to keep the wing flat. The first binding turn should be just firm enough to slope the wing then each succeeding turn tighter and towards the hook eye.

Love's Lure

A reliable pattern and credited to some unknown and probably long gone New Zealander, Love's Lure was probably created to imitate the odd truefly, bluebottle or housefly that ends up in the river, or maybe a bashed-about beetle. That versatile feather, peacock herl, is the mainstay.

Hook	12 to 16
Thread	Black
Body	Bronze peacock herl
Wing	Green peacock sword
Hackle	Black

With this pattern it needs a little practice to tie the overwing low and sloping. The hackle is tied full but also sloping back. For the dry fly version, it is the same dressing, but using a very good quality cock hackle.

NYMPHS

Hare and Copper

One of the most popular nymphs used in New Zealand and it looks like being with us forever. I think the reason is that the material is so easy to obtain and the pattern not too difficult to tie. To the trout it probably looks very 'buggy' and it can hide quite a bit of lead wire. Very few tiers keep to the standard pattern and I am no exception. My Blue and Grey Darters are really refinements to the H & C.

Hook	10 to 16
Thread	Black or brown
Tail	Tuft of hare fur (guard hair)
Body	Hare fur ribbed with copper wire
Over Thorax	Tuft of guard hair

Take the thread to the bend and tie in the tail tuft. Wind the wire in close wraps to the tail and then use it later to rib the fur body. The dubbing should be a mixture of fawn underfur and darker upper fur. Thicken the nymph up to the thorax and include some longer guard hairs. On completion, pick some of these out to simulate legs and make the nymph 'fuzzy'. Do not overdress the thorax tuft.

Pheasant Tail

This hardly needs description but some background may prove helpful. Most of our small nymphs are brown and using the russet/red strands of cock pheasant tail imitates more than one species. As with the H & C there are a few variations. The original used only fibres or herls tied using fine copper wire only, other patterns provide a wing case of grey mallard.

Hook	12 to 16
Thread	Black or brown
Tail	Pheasant herl
Body	Pheasant herl

Long tail fibres are best as the nymph can be tied from tail to wing case in one operation. After the tails are in the body is formed (tapered), the fibres tied behind the thicker thorax, then folded back to the eye to complete the nympal wing case. A fine wire ribbing is recommended.

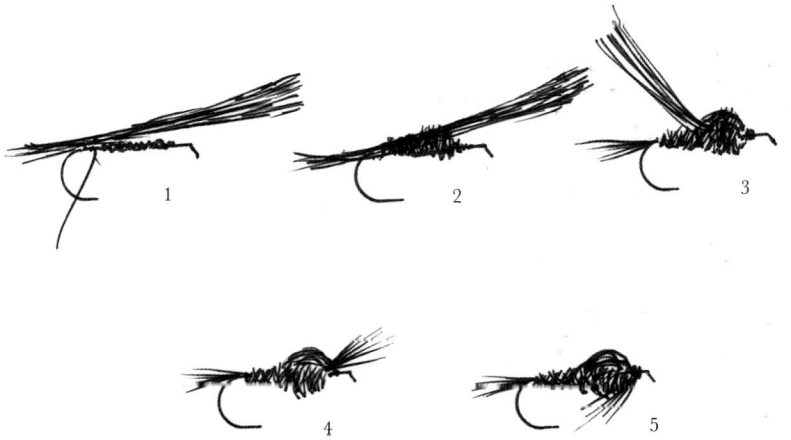

1. Tail fibres tied in. 2. Body formed. 3. Wingcase fibres pulled back. 4. Wingcase fibres pulled forward. 5. Fibre tips pulled under. (Item 5. For simulated legs two bunches of fibres are necessary.)

Possum & Peacock

A typical New Zealand pattern using materials readily to hand. As most backcountry guides will agree the P & P is a very reliable nymph when confronted with large wilderness trout. Even in the larger sizes it doesn't seem to alarm them, no doubt they regard it as a treat.

Greenwells Glory. Note proportions and neat eye.

Coch-y-bondhu (variation). Note substantial gold tag and body. Hackle-red and grizzle.

Partridge & Orange. Note streamlined shape and hackle length.

Grouse and Purple. Note well spaced rib and flared hackle.

Pheasant Tail. Note picked out thorax and substantial wingcase.

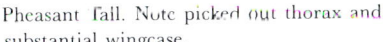

Blue Darter. Note very rough outline and darker thorax.

Water Boatman. Note substantial wingcase and heavy dubbing.

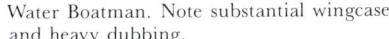

Passion Fruit Hopper. Note sparse hackle.

Hook	10 to 14
Thread	Brown or black
Tail	Pheasant tail fibres (short)
Body	Ruddy/red opossum fur
Thorax	Black opossum fur/hare guard fur mix
Wingcase	Bronze peacock herl

A big brown may not take it, but he will not ignore this bulky nymph. Last season, fishing a mountain stream, not 100 kilometres from Nelson it convinced a very large trout.

'Look at that!' I muttered, spying a fish just upstream of a boulder the size of a small shed.

'That's a b----y good one,' said Ian in his typical colourful language.

It *was* a big fish and sometimes big fish can give you the wobbles. Any fly fisher who says different — is either made of steel or a bit blank. On Ian's insistence I slid down the bank and unhooked the fly. It was a difficult fish to reach but eventually the P & P plinked in a metre or so upstream. I held my breath as the nymph drifted down.

'He's got it!' shouted Ian from his vantage point in the bush and the hook went home.

In dismay I watched the line disappear, but when the backing appeared and the big trout kept going I was ready to throw in the towel. But it must have been 'happy hour' for the fishing gods because, just before the rapids, the fish turned and came upstream. After what was minutes but seemed like hours, Ian was ready and a near 4kg brown slid over the net rim.

Our encounter wasn't quite finished. The fish, somewhat tired, rested against a large stone, then, to our surprise, a very large eel appeared. It nestled alongside the groggy warrior with the trout apparently unaware of the danger.

'You b-----d!' shouted Ian, and hurled a small boulder at the eel which shot downstream, the trout tearing off in the other direction. It was good riddance and a better ending. I know that eels have a legitimate right in the river's ecosystem but I'm damned if they are entitled to my returned trout.

'All comes to those that wait'.

11

Creative Patterns

There are times during a mass hatching period of an aquatic insect species that the trout will feed on them almost exclusively. I have had my share of those frustrating days when, despite rummaging through the fly box and fishing as carefully as I can, my efforts have been ignored. Many other fly fishers have shared this experience but have used their observations and creative skills to overcome the problem. For the non-tyer there are some shop-bought artificials that are good imitations of natural insects but it is left to the ingenuity of the fly tyer when the fishing really gets tough.

Many of the tying methods shown in the following patterns are based on New Zealand fishing conditions and are the result of much experiment. The natural models are all insects that inhabit our rivers, streams and lakes and the aim has been to stress some marked character of the insect. A factor kept in mind when devising them, apart from size, shape, colour, has been 'action'. A trout fly which can copy an item of trout food using the first three will most likely catch a trout's attention and, if the angler is skilful enough to include the fourth, when appropriate, success will follow. Catching a trout on any fly is a pleasure and if it is home-tied even more so. But catching a trout on one's own creation based on observation is better still.

Blue Darter

This nymph imitation takes its name from the rich blue wing case and is based on *Nesameletus* mayfly, a common nymph noted for its rapid swimming motion. The forerunner of the 'blue' appeared in my earlier book, *Trout Stream Insects*, but with a grey mallard wingcase. Both are successful trout catchers but I find the darker wing case has given even better results. If confidence is a major factor in angling success then by all means try this artificial. It rarely lets me down.

Hook	10/14
Thread	Brown
Tail	Hare fur tuft (fawn)
Rib	Fine gold thread
Body	Hair fur (fawn)
Thorax	Blue/black hare underfur/fawn guard hair
Wingcase	Mallard wing (blue)

Tying method

(The fawn guard tail hairs are taken from the back of the hare skin where the underfur is blue/black. Cut deep just above the grey skin hair. Keep this hair bunch pinched between thumb and forefinger and tease out excess fur leaving a small tuft of fawn guard hair. The excess dark fur is stored for later use as thorax material. If required a lead base is wound prior to tying (this applies also to other patterns).

1. Run thread to bend and tie in gold thread. Use loose wrap method to settle tuft or guard hair on top of shank. Before securing, pull hair butts gently towards eye, leaving the tail roughly half the length of the shank. Save excess dark underfur for thorax dubbing.

2. Dub fur to well-waxed thread sparingly and, after twirling the fur to a slender cigar shape, wind to behind eye and return to thorax area.

3. Follow with gold thread in three spaced turns to thorax and secure.

4. At the rear of thorax area tie in blue wing case slip with two turns. Butts to the rear with blue side under. When preparing wing slip (8-10mm) try to keep the fibres from splitting. Use remaining dubbing to form substantial thorax returning thread to behind eye.

5. Pull slip over top and sides of thorax tightly before securing with at least four turns of thread. Use thumbnail to press the wing case hump back, leaving ample room to tie off head. Pull tight again and take another four or five turns using these to control any rogue thorax hair. Cement head threads by taking one drop of cement on the point of stiletto, deposit this on eye threads and stroke the remainder underneath.

Thorax dubbing should contrast well with abdomen dubbing. Use excess fur from tail material plus more dark underfur with some guard hairs included. Guard hairs will dub successfully when mixed with softer fur.

This is mixed ready for dubbing by placing tufts in the palm of the hand, rubbing it gently with the forefinger, and teasing out lumpy fur.

6. With stilleto pick out short tufts of dark fur from thorax, laterally and downwards, making sure that a few guard hairs are also prominent to suggest legs.

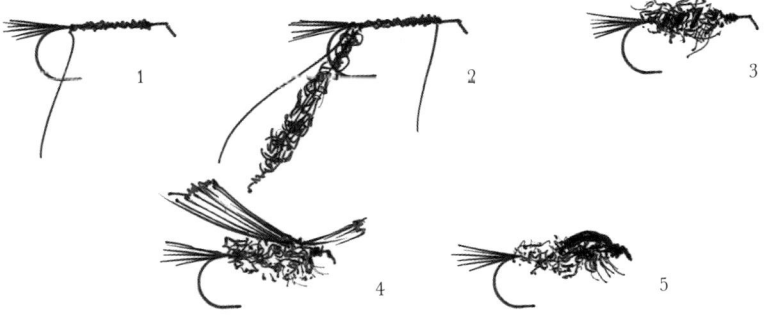

1. Tail tied in. 2. Body materials tied in. 3. Body formed. 4. Wingcase tied in. 5. Wingcase folded forward and secured.

Woolly Caddis

Simple to tie, the woolly caddis is a fine trout catcher. A copy of this uncased caddis was the pattern I created many years ago and, since then, it has seduced many trout. It is most useful during the early spring when rivers flow higher than normal. Normally fished weighted it hugs the stream bed where drifting woolly caddis are common. Some rivers host more woolly caddis than others. The Arnold flowing out of Lake Brunner is one of them. While in the area, and with a day to spare, I checked out the Arnold in the vicinity of the hydro dam, but the deep flowing water, while ideal for spinning, proved unsuitable for my style of fly fishing. I moved upriver and just below the bridge found open water which looked promising if a little high. A few casts with a dry Pheasant Tail introduced me to a very lively Arnold fish of around 1kg and it was killed for the next day's meal. On opening the fish I found it gorged with woolly caddis.

So, despite an early catch, I tied on my imitation. It doesn't work all the time but the next few hours convinced me I had made the right choice. The Arnold is a fast stream and getting the caddis to the bottom was only achieved by adding another weighted nymph to the tippet. It may be fast but the Arnold, if slightly stained, is reasonably clear, and by fishing near the bank edge it was easy to see the flash of a trout turning to the nymph. By late afternoon, after returning many gold flanked trout, I was well satisfied. Cobble bed streams are ideal for this grey caddis and trout are well aware of its presence.

Suspender Midge. Note Ethafoam ball tied forward.

Woolly Caddis. Note rough dubbing and large thorax area.

Turkey Tail nymph. Note swollen wingcase.

Kakahi Queen Nymph. Note thick abdomen.

Brown Stonefly. Note body segments.

Latex Caddis. Note, latex can be coloured with marker pen.

Starling Emerger. Note Sparkle Yarn or similar mixed with fur dubbing.

Creedon's Creeper. Note thick dubbing and simulated legs.

Pattern

Hook

18/20 Caddis 37160 or Partridge Grub

Thread	Black
Tail	Hare fur tuft (fawn)
Rib	Brown varnished wire (fine)
Body	Hare/rabbit fur (grey) mixed with black tipped guard hair
Thorax	Black floss silk

(Note). Mustad Caddis Hooks have higher hook numbers than standard ie.
Standard 94840 Caddis 37160

14	20
12	18

Tying method

1. Run thread to end and tie in short bunch (10mm) of brown hen hackle fibres. This should be tied in at the extreme bend of hook so that the fibres are pointing slightly in the direction of the eye. Note: This is made easier by temporarily turning the hook upside down.
2. Tie in wire close to fibre butts.
3. Form the rough *stout* body with fur dubbing only *slightly* tapered to eye. Note: the fur dubbing is similar to that used for the Blue Darter but mixed a darker tone. Be generous with this dubbing as the wire tends to suppress the fur.
4. Tie in floss silk behind the thorax area. Note: Floss silk is easily ragged so smooth fingers are necessary. For a substantial wing case double the silk then double again but do not fold at this stage.
5. Form a substantial thorax with additional dubbing, pull the four strands of silk over and secure. Note: Keep the strands touching so as not to expose back dubbing.
6. Pick out guard hairs from eye to bend forming a fringe along the sides and belly. With scissors cut excess fur from curved back of the caddis.
7. Smear the wing case liberally with cement to form a black shiny head.

Turkey Tail

An inspection of *Deleatidium* nymphs will show there is a recognisable colour difference between the juvenile nymphs and those at the hatching stage. The former are a pale or russet brown and the latter a rich umber especially the wing pads. Sawyer, that now famous river keeper of the River Test, was inspired to imitate the immature nymph with his pattern, the Pheasant Tail. My pattern is a variation of the Pheasant Tail but closer to an imitation of a hatching nymph.

Hook   10/18 Mustad 94840 or 9671 1XL

Thread	Black
Tail	Fibres from turkey tail feather
Rib	Fine gold thread
Body	Turkey tail fibres
Thorax	Black/blue hare fur
Wing case	Turkey herl

Tying method

1. Run thread to bend and tie in bunch of fibres (6-8) from white tipped turkey tail feather but do not use the white-tipped fibres. Note: Turkey tail fibres can be up to 8cm and, unlike some pheasant tail fibres, provide ample length. The tails should extend shank length.
2. Tie in gold thread at bend.
3. Twist and wind fibres to behind eye and return fibres and thread to behind thorax area. Note: Start abdomen with butting turns but thicken the body as it approaches the thorax area.
4. Wind gold thread to thorax area and secure. Note: Three turns are enough. The fine gold thread will not crush herl.
5. Dub dark hare fur and form a plump thorax. Note: Ensure a few guard hairs are mixed with thorax dubbing.
6. Bring turkey fibres over thorax and secure. Return thread to rear of thorax and repeat overlay of fibres. Note: When stretched, the turkey herl quills pull through to the surface, leaving a very realistic hard, semi-shiny wing pad.

Hopper

Distinct from the grasshoppers, the passion fruit hopper is a small bug, a sucking insect that during the late summer months attacks soft fruit crops. The juvenile hoppers appear as tiny pieces of white fluff but it is the adult that becomes a seasonal trout food insect and is easily recognised by its triangular wing shape and its remarkable jumping ability. It is also easy to imitate, but keeping the wings flat against the body has always plagued the tyer. I solved this with a special tying method.

Hook	14/16 Mustad 94840
Thread	Brown
Wings	Cock pheasant (church window back feather)
Body	Opossum belly fur (fox-red) /Hi Vis mix
Hackle	Ginger cock (small)

Tying method

1. Run thread to bend and tie in wing. This is especially cut to shape. Strip a large church window feather (75mm) leaving tip of feather (20mm). From this cut fibres at right angles up to quill but without cutting it.

This leaves the delta shape which will be approximately 8mm all sides (see sketch). The protruding fibres are tied in at the bend with quill free and to rear.

2. Dub fur/hi-vis mix onto well-waxed thread and form plump body.
3. Tie in small (30mm) hackle near eye.
4. Take wing quill and pull over body. The rear anchoring fibres will ensure the wing lies flat. Tie quill down behind hackle securely.
5. Wind hackle (3 turns) in front of wing and secure.
6. Take up SMALL drop of cement on stiletto point, smear onto flat wings and stroke gently.

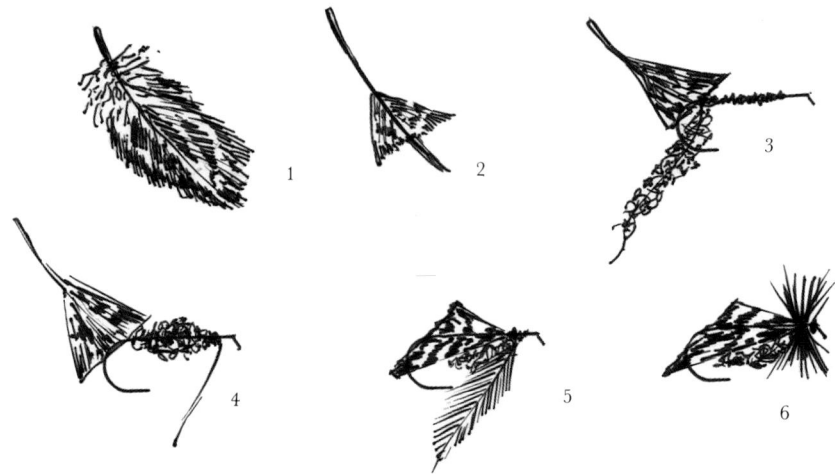

1. Pheasant back feather. 2. Shaped feather. 3. Wings and fur dubbing tied in. 4. Dubbed body formed.
5. Wings folded forward and hackle tied in. 6. Hackle wound.

Kakahi Queen Nymph

Modelled on a mature *Coloburiscus humeralis* nymph, the main characteristic is the spiny gills over the abdomen, the overall colour is a sepia brown. Apart from the most sluggish of waters or lakes this mayfly is common but especially in well-oxygenated rivers. The use of peacock herl to imitate the gill area once again shows the versatility of this feather. The burgundy of red/brown silk, when wet, turns to a dark chocolate colour.

Hook	12/16 Mustad 3666 or similar
Thread	Brown
Tail	Short bunch ginger hen (8mm)
Body	Bronze peacock herl
Hackle	Ginger hen (30mm)
Thorax	Burgundy or red/brown floss silk
Wing case	Goose quill slip (dyed brown) or substitute

Tying method

1. Wind thread to bend and tie in hen fibre tail (10mm).
2. Tie in length of floss silk and peacock herl (three strands). Note: Choose long bronze fibre herls. Return thread to thorax area.
3. Wind silk to eye, tie in hen hackle and return silk to thorax.
4. Twist herl and wind to thorax in even turns without overlap and secure.
5. Tie in generous section of goose quill feather mid shank.
6. Form plump thorax with floss silk and return thread to behind thorax.
7. Wind hackle (three turns), pull fibres down, sloping to bend, and secure. Note: Hackle should be kept short (10mm).
8. Hold wing case evenly over thorax and secure. Note: The wing case fibres should form a slight shoulder around the thorax silk.

Brown Stonefly

The green stonefly is the more colourful but the brown stonefly is more common. Throughout the season, apart from the most sluggish of waters, trout feed on stoneflies and they are well worth imitating. Strong features of these very robust insects are the stout tails, the somewhat flat abdomen, and the prominent wing cases. These characteristics are emphasised in the following pattern:

Hook	8/14 Mustad 9672 2XL
Thread	Brown
Tails	Brown goose biots
Underbody	Cream or amber floss silk
Wrap	Sheet latex
Overbody	Brown goose quill fibres
Thorax	Opossum red/brown belly fur/hare guard hair mix
Wing cases	Mallard dark grey wing slips. (pre varnished)

Tying method

1. Tie in lead wire. Note: This is described because of the technique. Take thread one third from eye. Tie in two short lengths of lead wire on either side of shank and tie in to near tail. Take thread back towards eye and overtie lead.
2. Tie in paired goose biots (10-15mm) at bend. Note: The fibres should be torn from the shorter fibres of the quill feather.
3. Tie in narrow strip of latex (5mm); underbody of silk and feather overbody. Note: Cut the latex strip to a point before tying in. This will help form taper.
4. Wrap silk over the lead making sure there are no gaps, and secure. Cement the overbody feather, pull firmly over silk and secure.
5. Wrap over silk and feather with latex and secure. Note: The wraps should overlay slightly to form a realistic segment effect.

6. Tie in the wing case strip which should be wider than the finished abdomen (5-8mm) and form the substantial thorax in front.
7. Press the wing strip back, crimp with thumbnail, and fold forward. Secure this mid thorax and repeat the dubbing/wing case operation once more.
8. Pick out dubbing and guard hairs to simulate legs.

Cased Caddis

Cased caddis are a favourite trout food on most rivers and lake edges. Almost without fail a trout autopsy will reveal scores of these curved larvae. I have found up to a thousand of these grubs packed in some particular trout stomachs. Only during floods and freshes do trout take them on or near the surface. This is when the tiny insects cannot retain their usually tenacious hold on the stream bed and are at the mercy of natural buoyancy. However, I have seen trout grazing them from rocks. Imitations are best weighted and tied on caddis or grub hooks to copy the shape of the natural.

Hook	16/20 Mustad Caddis 37160 or Partridge Grub K4A
Thread	Green
Body	Green Swannundaze (#19)
Thorax	Bronze peacock herl/rabbit fur mix

Tying method

1. Take thread to around hook bend and tie in plastic lace. Note: Before tying in use scalpel to sharpen end of plastic. If necessary the plastic can be thinned by stretching in hot water.
2. Wind lace two thirds to eye and secure. Note: When wrapping keep the lace tight and all wraps butting to form segmented effect.
3. Dub thread with rabbit fur and together with one strand of herl form bulky thorax. Lay herl back over thorax to swanundaze and secure. Fold herl again back to eye and secure. Note: This will force the fur outwards to simulate legs. Clip excess herl from underside of thorax.

Starling Emerger

All credit to Vern Williams, my fishing buddy, for this great trout catcher. It probably takes more selective trout than any other — those very choosy trout feeding in slick glassy runs. In Vern's words, 'The starling hackle is very soft and mobile and creates the illusion of "life". In trying to rid itself of its nymphal case the emerging dun flexes and the legs move which makes the starling feather an ideal choice for hackle. This creates a "trigger" reaction in the trout.'

Hook	14-16
Thread	Brown
Wing	Brown polywing

Tail	Bronze mallard (breast)
Body	Rabbit/red opossum mix
Hackle	Starling

Tying method
1. Tie in a tuft of fawn polywing, upright, and clip short.
2. Tie in prepared hackle behind wing tuft, take thread to rear and tie in three short speckled mallard fibres.
3. Form dubbed body to behind wing tuft.
4. Wind hackle with two turns of the soft fluffy strands behind the tuft and three turns of the shiny fibres in front which help support the fly in the surface film.

Creedon's Creeper

Al Creedon of Wellington is a creative tyer. His creeper has proved successful in many of our rivers and streams. It is a cleverly constructed trout fly that imitates more than one trout food insect from dobsonfly larva to Dragonfly nymph and uses readily obtainable material. An unusual feature is that the fly is tied starting from mid-shank.

Hook	Mustad 3666 #10-12
Tail	Cock pheasant tail (3)
Body	Chenille dark brown
Wing case	Hen pheasant wing strip
Legs	Cock pheasant tail fibres
Thorax	Peacock herl

Tying method
1. Start mid-shank and wind thread to bend. Reverse thread a few turns and tie in tail fibres with tips to eye. Double each fibre back in turn and form separated and spread tails.
2. Return thread to mid-shank, tie in length of brown cotton, twist cotton and chenille into rope and form plump abdomen between bend and mid-shank. Cut excess.
3. Tie in wing case and cut excess but do not fold. Tear or pluck three cock pheasant fibres from quill stem which will leave simulated claws on each fibre and tie farside mid-shank with fibre tips pointing and level with bend.
4. Fold remaining claw fibres to rear. Repeat on near side. Tie in herl strands and form plump thorax and remove excess at mid-shank.
5. Bring the far side pheasant tail claw fibres forward and secure mid thorax. Repeat near side fibres. Collect the wing case strip, pull forward over thorax herl and legs and secure forming neat eye.

It was thanks to Trout Unlimited that I discovered the effectiveness of Al Creedon's pattern. Invited to speak at a New Plymouth seminar we had parked the van alongside a nearby lake and when I spotted a large brown cruising the lake edge I knew we were in business. I also noticed he wasn't rising but was searching the weed beds. Searching my fly box for a good imitation of a dragonfly nymph I came across the Creeper, a gift from artist/fly tyer John Morton. Inside a couple of hours that fly had accounted for three browns, none less than 2kg. That is after I had nipped a tiny lead shot to the tippet. It might be tricky to tie but it certainly tricks trout.

Trout flies are always in a state of evolution, thanks to the innovative tyers. But, despite some wonderful imitative patterns that look so real you feel like swatting them, the student tyer can sometimes take the cake. I recall one Friday morning when a local shopkeeper phoned and asked me to guide an American and his son for a day's fishing on the Motueka. I'm not really a guide but they had had little luck on their travels so despite the short notice I agreed. Jim, the father was a most friendly fellow and Ben, the son, one of the most respectful boys I have come across. Their tackle was well up-to-date and it was obvious they were no novices.

By 10am we were tackling up at the tail and one of my favourite pools. I pointed out three good fish, none rising but active and obviously feeding. I profess a reasonable pride in my fly box contents but when Ben opened his I slipped mine back in my pocket. His skill was extraordinary. Mayfly and stonefly nymphs in rich colour and all with that 'buggy' look that you know will convince fish. And his caddis imitations were a delight to see. Jim's box was pretty rough and I was eager to see how each would fare. Jim took the first fish, a nice two-pounder lying along the shallow edge, followed by Ben who, despite some nifty casting, put down the next one. That proved to be the pattern for the day with Jim outfishing Ben by at least 3 to 1. That performance confirmed my belief that trouts' eyes are the best judge of what they like to eat and we can put too much emphasis on 'arty' trout flies. Thank goodness trout enjoy 'rough' flies otherwise many of us would have to sing for our supper!

Hard to reach but worth it.

12

Yankee Doodle Dandies

New Zealanders can hold their own in any fly fishing stakes but I think we tend to become a bit parochial in our choice of trout flies. In recent years, no doubt owing to the increasing number of overseas anglers, particularly American, our territory has been invaded. Some overseas patterns are no better or worse than our run-of-the-mill flies; others are simply fancy flies. But some are not only a tribute to the fly tyers' skill and ingenuity but are also great trout catchers. The following patterns have earned a place in my fly box.

Henryville Special

First introduced to this adult caddis imitation by Bob Gormley of Rhode Island I have since enjoyed marvellous success with it during the caddis hatches. The Henryville Special, was created by Hiram Probst of Pennsylvania U.S.A. On many trout streams juvenile caddis are the most common trout food insect. But the adults are just as plentiful and, maybe because hatches usually occur in late evening, they attract less attention from the fly fisher. When a caddis hatch is in full swing it is near impossible to catch a trout on any other floating imitation and a good copy of a natural adult caddis should be a priority for the fly box. In the 'Special', Probst

117

incorporates the key characteristics of the caddis fly — the low-swept wings, slim fuzzy body with the additional underwings emphasising the shape of the natural insect.

Hook	94840 #16/18 1XL
Thread	Brown
Palmer hackle	Short fibre grizzly
Body	Light brown possum fur/grey rabbit mix
Underwing	Bronze mallard breast
Overwing	Grey mallard wing slips
Throat hackle	Short fibre brown cock

Tying method

1. Form thread base from well behind eye to bend and tie in grizzly hackle, butt first.
2. Form slim body (no need for taper) and palmer (wide turns) to thorax area and secure. Trim hackle fibres from top of hook and tie in underwing (not thick) to nice slope.
3. Pair the mallard wing slips and tie in concave (hollow side inwards) with tips to rear and upwards, the lower edge of slips parallel to hook shank.
4. Tie in and wind brown throat hackle.

   One of the best caddis hatches I have ever seen occurred on the lower reaches of the Buller River. During the day, using a creeper imitation fished through the cobbled shallower stretches, I landed some decent size browns and a lovely cruiser which took a #16 Dad's Favourite. Trutta, my faithful labrador, spent some time chasing 'flappers' with an anxious mother keeping a watchful eye on the dog's antics. After dinner, within sight of the campervan, I returned to the nearest pool just in time to see the first splashy rise. Wading knee deep just above a large riffle I floated a Henryville caddis down the chuckling water and was rewarded with a strong pull, followed by the sound of a whirring reel. The hook held (not always so when fishing downstream and across) and a fat brown graced the net. Others followed but were only tempted by using a little ploy of twitching the fly upstream before it had time to drag. Around 9pm the first hint of a weather change arrived with a cool downstream breeze and, as if by magic, the caddis disappeared. The end to my fishing was when I felt something nudge my leg and in the gathering gloom I saw a monster eel. I hardly touched the surface on my way out!

Humpy

The name describes the fly. Unlike the Henryville Special this pattern bears little real resemblance to any natural but owes its success to its special floating qualities and the very original tying method. It is thought the designer of the Humpy prototype was a Jack Horner of the U.S.A. No need for a wide range of materials here. Just deer hair, and a bit of silk hackle, the former providing tail, body and wings.

Hook	#10-16 Mustad 9672 1XL
Thread	Brown
Tail/Overbody/Wings	Deer hair (fallow or Virginia)
Underbody	Yellow, green or red floss silk
Hackle	Brown cock

Tying method

Start tying thread mid shank and tie in tails returning thread to mid shank. Clip hair butts at this point and prepare another thin bunch of hair the tips of which should extend beyond tails a half shank length. All hair butts are now mid-shank and are bound in common turns, the thread now at the bend. Tie in floss silk, return thread to just over mid-shank. Wrap silk to mid-shank and secure silk. Separate tail and body fibres and bring the latter over the silk and secure just over mid-shank. Lift the body hair tips to vertical with thread turns at the wing base and split hair using figure eight turns. Finish with hackle and eye turns. High quality cape needs one hackle only, others two.

I cannot recommend the Humpy too highly. Alan and I were fishing a backcountry river, not far from Lake Mavora and, although most of the fish were not too fussy, one in particular must have had a degree in selecting trout flies. I'd spent well over a half hour on the fish which was lying deep in fast water and naturally plied him with bymphs of all sizes and many patterns. Sometimes he'd slide sideways for a look or lift a little off the bottom but that was all. 'Try a dry' said Alan. 'He's pretty deep', 'You try him,' I replied, fairly confident that he was too far down. His Humpy bounced down the rapids between two large boulders and over the big fish which, as if he hadn't had a feed for weeks, lifted to take the fly. Which proves no matter how long you've fished for trout you're always learning.

Muddler Minnow

This is a perfect example of creative fly tying. Credited to Don Gapen, a talented American fly tyer, the Muddler seems to be a chameleon of trout flies in that it can imitate a small bully, cricket, cicada, or any other chunky natural fly. The brown natural colours and gold tinsel are its attractive features.

Hook	Mustad 94840 #10/12 2XL
Thread	Brown
Tail	Turkey wing fibres (mottled)
Body	Flat gold tinsel

Underwing	Brown squirrel
Wing	Mottled turkey
Head Collar	Deer hair (fallow)

Tying method

1. Tie in short strip (half shank length) of turkey tail fibres at bend but thread wrap remainder to shoulder.
2. Tie in tinsel, wrap to tail and back to shoulder.
3. Tie underwing on nice slope with fibres surrounding shank (slack turn first then tighter) tips at mid-tail. Cut excess butts.
4. Pair turkey wing slips and tie in on slope with tip edge up or down (optional) but tips to mid tail.
5. Tie in small clumps of deer hair, spin and trim to form small bullet head.

My first experience with the Muddler occurred when fishing a small Taupo tributary and, oddly enough, it took a famous English angler to introduce me to this American trout catcher. The rainbows were running with each pool holding up to four fish all vying for the best position. Before the day was out, between us, we had landed many large rainbows, keeping a couple for the smoker. The Muddler worked so well it earned a permanent place in my fly box and has proved itself on many more occasions. Despite our success fishing for rainbows spawning it is not my cup of tea. I find that real angling enjoyment comes from landing fish that are hard to catch. Cream on your cake is nice but you can have too much of it.

'Which fly'?.

Irresistible

One of my favourite patterns when fishing backcountry rivers and in small sizes on small streams. It is a wonderful floater and perhaps because of the sharp contrast between fore and aft, easily spotted by trout in rougher water. I can picture the small river quite clearly. Dominated by snowcapped mountains it ran through a great tussock plain and was home to some very large rainbows and a few mouth watering browns. New to the river I was delighted when my two companions, generous to a fault, put me behind fish after fish and 'praise the lord', I hooked some of them. 'Hooked', is the word. 'Landed', not too often, in that fast running ice cold stream. In every case the running rainbows stripped the reel to near the spool and the fly? The Irresistible.

Hook	Mustad 94840 #10/12/14
Thread	Brown
Tail	Short tuft of fine black tipped hair. (Squirrel tail)
Body	Deer hair
Wing	Squirrel tail
Hackle	Dark blue dun (dyed grizzle hackle) and dark brown hackle.

Tying method

1. Using a cut-off cartridge case tap a tuft of hair (tips down) in the shell.
2. Tie in tail (shank length).
3. Spin deer hair then trim forming stout body.
4. Tie in wing tuft and raise to slightly forward. (tight turns)
5. Tie in hackles.
6. Wind to form bushy collar over clipped wing butts.

Adams

The creation of Michigan angler, Leonard Halladay, it is the most popular fly used by American visitors and is becoming increasingly popular with New Zealand fly fishers. To a trout, the grey body and barred wings may well give the impression of some of our lighter-winged, newly-hatched duns.

Hook	14/16
Thread	Black
Wings	Grizzle hackle tips (hen)
Tail	Grizzle and brown (cock)
Body	Grey rabbit fur
Hackle	Grizzle and brown

Tying method

1. Select two hackle tips after stripping away most of the fibres (8mm for

#14). Tie upright using figure eight wrap to divide them slightly.

2. Tie in bicoloured tails and split with one turn through.

3. Form tapered body with dubbed fur and tie in the two hackles. If one is slightly shorter than the other then tie the longer one in first. Three turns behind the wing, one turn in front.

Royal Coachman

We tend to think of American flies as modern but the Royal Coachman was first devised in 1878 by John Haily, a professional fly tyer from New York. It is really an offspring from an old English pattern with added title. In more recent years it has another relation, the Royal Wulff, named after its creator, the late Lee Wulff, remembered as one of America's noted fly fishers.

Hook	10 to 16
Thread	Black
Wing	White polyyarn or calf tail
Tail	Golden pheasant tippets
Body	Bronze peacock herl with red floss mid-rib
Hackle	Dark brown

Tying method

1. Tie in polyyarn, or calftail, split and secure upright.
2. Take thread to bend and tie in tippets, herl, and floss silk.
3. Wrap floss just short of mid-shank and follow with herl just beyond mid-shank.
4. Trap herl with floss and form short floss body over herl.
5. Carry on with herl to behind wings and secure.
6. Wind hackle (usually two) and secure.

A very satisfying fly to tie and to the tyer's eye, rather beautiful. The crisp white wings set off the sparkling herl while the red cummerbund adds just that sporty dash. The trout must agree because it dupes a lot of them.

Compara-dun

From the trout's viewpoint, these sparsely tied floaters must seem very realistic. It is very successful, especially on slow-flowing stretches of river. Compara-duns are ideal for fishing on still ponds, their special feature being the elimination of the traditional collar hackle. They were popularised by Americans, Al Caucci and Bob Nastasi, but the concept goes back to the 1930s. Hackle collars have long been a discussion point between fly fishers. An underwater view will explain why. The myriad of hackle points seem nothing like the legs of a natural fly but are necessary to float the fly. Compara-duns use only a hemisphere of fine hair which allows the hair fibres to straddle the surface film without piercing. Rear support comes from strong, widely spread tails.

Hook	14 to 18
Thread	Black
Wing	Fine whitetail deer hair
Tail	Dark ginger hackle fibres or synthetic Micro fibbets
Body	Red brown opossum fur

1. Wrap a firm bed of thread, tie in a sparse tuft of hair, then gently push the hair fibres down and around the *sides* of the hook with the left thumb.
2. After lifting the fibres upright, secure and adjust them with tight turns in front. They should now appear as a fan over the hook.
3. Cut the excess hair on the slope, take thread to tail and form a small ball of fur near bend. This splits the tail fibres and creates tiny outriggers.
4. Form the tapered body but carry it past the wings to near the eye. This extra fur forms a nice thorax and helps support the spread wing hair.

Greenwell's Parachute

Whoever thought of tying a hackle horizontally was a master of lateral (excuse the pun) thinking. The placing of the hackle gives maximum support to the fly without unduly breaking the surface film. The problem is that you need something to wrap it around which means some sort of stem. Some hooks are designed with this feature but a wing base of some sort is normally used. Similar to compara-duns, they present the trout with a natural impression of the floating insect.

Hook	12 to 18
Thread	Black or brown
Wing	Grey Polywing (clump)
Tail	Furnace hackle fibres
Rib	Gold thread
Body	Olive fur (dyed rabbit)
Hackle	Furnace

A straightforward tie until you reach the horizontal hackling, an unusual technique but one that is quickly mastered with a bit of practice. I find it easier to use a home-made gallows suspended above the vice. To this I attach the small spring from a spent Biro and a small electronic spring gripper. This holds the wing clump rigid while the hackle is wound. Good quality hackle is essential. Long, short fibred, feathers are ideal.

1. Strip lower fibres from one or two top quality hackles depending on fly size, leaving a long stalk.
2. Tie in at wing point with hackle butt to rear and form a small loop with stalk.
3. Complete tails and body and pull wing upright using the gallows tool.
4. Wind hackle *down* the stalk (each turn *must* be below the preceding one).
5. Slide thin-nosed tweezers through the stalk loop and pick up the hackle

Henryville Special.

Humpy.

Muddler Minnow.

Irresistible.

Adams.

Royal Coachman.

Compara-dun.

Parachute.

tip. Retrieve hackle tip through stalk loop and pull tight towards the eye. This in effect ties the hackle in a knot.

6. Complete head whip under the horizontal hackle. This is where a whip finishing tool is ideal.

For the Greenwells pattern a stub of grey Polywing is used, with the hackle simply wound as shown, and made off underneath the hackle close to the stub and on the eye side.

Norman and Jean Marsh in Fiordland.

13

Tactics

Whatever pundits may say, it is a fallacy that fly fishing is easy. Maybe sometimes the trout *can* be easily fooled but not too often. It takes skill to cast a fly and even more to deceive a trout. Luck, as any angler knows, plays a significant part in successful fly fishing for trout but as the years pass and experience grows the odds increase in the angler's favour. Much of this knowledge is passed on from fellow anglers.

It would have been nice to have had a fishing mentor in my early years, perhaps some kindly uncle who not only produced little fishing gadgets at Christmas but who could charm your mother out of whacking you on returning late from the river. No such luck. With no fishing relatives it was just me and the push bike. Before that, shank's pony. A substitute for that mythical uncle was the Nelson Public Library, which I hasten to add was in my home town in Lancashire. In that hushed atmosphere, enthralled, I roamed the world with a fishing rod, from the Rocky Mountains of Canada to the dazzling turquoise-blue flats of Florida. Sitting under the 'QUIET PLEASE' sign I battled great New Zealand rainbows, netted Norwegian salmon, or spent heavenly days on some classic chalk stream. Little did I envisage that one day my dreams would come true. But dreams don't catch trout. Experience and tactics do.

OBSERVING TROUT

Time spent observing trout is one of the best investments a fly fisher can make. Understanding the quarry is vital to any hunter and trout are no exception. They are endowed not only with keen eyesight but also with exceptional sensitivity. They cannot hear in our terms but become aware of movement through a series of sympathetic pores along the lateral lines. I recall fishing near road works where blasting operations were taking place and these were completely ignored by the trout. But a heavy crunch on the beach gravel is enough to alert them, if not send them scooting. But they do have an Achilles heel. An angler positioned in a narrow arc behind the fish, providing he is not less than 15 metres away, will appear indistinct and of very low profile. Keeping the rod low is obvious advice. An angler's approach from below is mostly the safest way but this is not always the case. In very low water conditions where the trout are super sensitive and where even fine nylon over their backs is often a recipe for disaster, a side-on approach, if not infallible, is much preferred. It may involve a cossack squat style approach over the beach but if humility is the price it should be readily paid.

When trout are resting or feeding very close to the surface, say along a beach edge, their range of vision is reduced to the extent that a crouched angler can approach to within a few metres. A fly or nymph cast just upstream of his nose will generally bring a response.

Casting to a trout from upstream is generally a recipe for disaster but there are times when you have little choice. It is possible to fool trout even under these situations but these are usually fish that are rising and, are more intent on feeding than 'spooky'. A rather long, fine leader is necessary and, with a minimum of false casting, the back cast is checked at 12 o'clock and the leader and tippet allowed to collapse a short distance upstream from the trout. The fly is the first object to appear before the fish and, if not 'dragging', is almost certain to create interest. Hooking the fish is a different matter as the fly is pulled away from the trout's jaws. As long as four seconds should be allowed before setting the hook.

There are trout and trout. Some are highly educated (in human terms); others, although maybe not as quick on the uptake, should not be regarded as village idiots. All living creatures are subject to the forces of supply and demand. The trout is no different. Come high water he will take advantage of the excess food supply and gorge. But those events are infrequent and, except for mass emergence times of insects, he survives on an irregular diet. We often come across a trout that seems asleep. Or is he just giving his stomach a rest? Trout in this somnolent mood can sometimes be induced or pestered into taking a nymph but that smacks of underarm bowling. A practical fly fisher moves on.

If alarmed, a trout may not move immediately but you can rest assured

he is now alert to danger. The reason for the delay, I believe, is that a trout's brain registers danger in proportion to shock. The sudden appearance of an object (i.e. an angler) will, to a greater or lesser degree, diminish the available surface light in the trout's 'window'. While a true explanation of the trout's vision is impossible many theories have been proposed, many calculations made, based on the incidence of light striking the water surface, natural laws. For instance we know that light rays are bent on entering water and that a trout position is an illusion and further away than we think. Conversely, from the trout's viewpoint, he may well see an image of a bankside figure displace from the real one. Regardless, I do know that trout have devilishly good eyesight and proved it the other day. While stalking gravel upstream shallows I spotted a trout rising occasionally some 20 metres distant. Bending, with rod trailing, I started to approach along the gravel edge at river level and to my surprise watched him sidle away to deeper water. Noting my position and a prominent upstream marker I paced the distance which measured near enough to 14 metres, say 40 feet. Even at shoulder height he must have seen me anxiously leaning over the edge of his 'window', that imaginary hunchback circular cone he is doomed to carry throughout his watery life. And no wonder. Owing to the 'bent light' phenomenon my image was well above my head, or at least the upper half of me. A bit indistinct perhaps but enough to ring his alarm bells. Surprising though because he wasn't deep and at that depth his 'cone' would have been much smaller, making my wedge of invisibility much in my favour. Perhaps that trout didn't understand the laws of physics.

There is one other nearly important target for the trout and that is the trout food insect whether on top or under the surface. I say nearly because, natural laws take preference and self preservation becomes paramount, almost to the starving point. But what does the trout see as he diligently waits and watches. Again, based on our own optics, we can only guess. Obviously he has keen eyesight being able to spot the tiniest food items in the often tumbling stream. Based on human optics, looking upwards from a stream bed through reasonably clear water, the underside of the surface has a mirror effect showing every detail of the stream bed and any item between.

If, in our imagination, we take the place of the trout this is what we may see. Looking upstream, the underside of the stream surface may have various items of small debris some completely submerged, others which have pierced the surface 'film'. The rest of the undersurface is the reflected image of the stream bed. When a small insect alights on the upstream surface it announces its presence through tiny starflashes of light as the tiny indentations bulge the surface producing watery light condensers. This provides an early warning system for the trout because the fly is too far away to give a surface view. As the fly approaches the edge of his 'window'

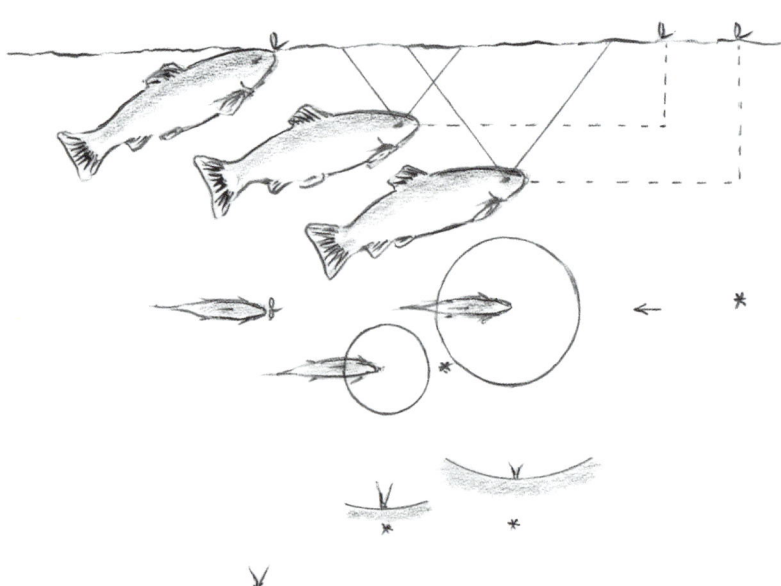

The trout's window.

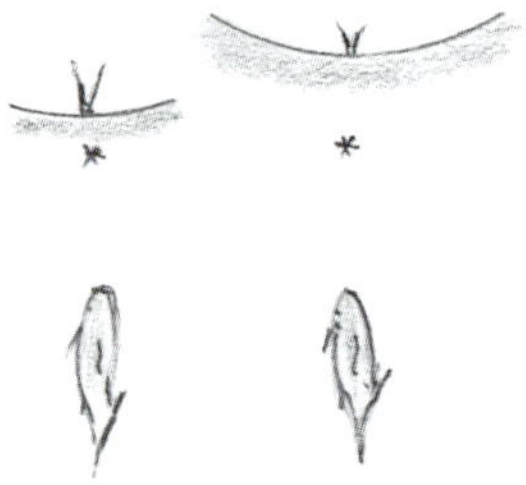

Easy to see.

the trout sees any submerged part of the body of the fly and, because of the 'bent stick' phenomenon, the tip of the wings. At the *edge* of the 'window', wings and body meet and the trout has a somewhat fuzzy view of the fly although it is in the best position to identify any colour.

Now comes the fascinating part. As the trout rises, his imaginary 'cone' is growing smaller and the projected image of the fly may seem to be sliding *down* a slope. At the surface the imaginary 'cone' has disappeared as has the false image which is exactly the trout's intention without all this malarky. What intrigues me is that just as we misplace the trout so does the trout misplace the fly which appears to be sitting above the surface, tilted and in fact head on. This means that, although we imagine he sees only the underside of the fly, he can see both body and wings. What does all this mind bending mean to the fly fisher? For one thing it means that wings on a trout fly may be important when trying to outwit selective trout and secondly, that pattern and colour may be more important than we think. It also suggests that a high riding imitation may give a better first impression. Better quality hackles?

What of a completely submerged wet fly or nymph? How does the trout see this? A submerged divided marker, yellow on top and red below will be reflected onto the 'mirror' showing, against the reflection of the stream bed, only the yellow side of the marker which will appear upside down. So will an imitation wet fly or nymph. In fact the trout will see two imitations, the real and the false. Which one will he choose? No problem. As the reflected nymph reaches the 'cone' edge it has moved out of the mirror range and disappears leaving the real one at the trout's mercy. Trout optics may be an intriguing subject but will it bring any more trout over the rim of the net? I go back to my opening experience. Next time I approach a trout, despite the fact I know he can probably see me I'm going to crouch a little lower, even with a crook back. I'm taking a bit more care in tying my trout flies and will continue to dress most of my dry flies with wings. With my average abilities I need all the 'edge' I can get.

Over a long learning period, a trout's instincts become tuned to various natural objects which appear on the rim of his circular 'window' but anything strange immediately spells trouble and he reacts accordingly.

I suggest that, even if his burglar alarm is buzzing, it may not be a full scale alert, and if left undisturbed for a period, he could well resume feeding. Only the other day, with a strong wind and ruffled water making it almost impossible to spot trout, I came across three good fish parked in a sheltered bay between willows. After a cautious approach the nearest of the three moved to the nymph but, on tightening, my only reward was a scale on the hook point. All the fish disappeared into the near-side current. Instead of giving them up for lost I stood patiently downstream for at least five minutes. A fish appeared, then another and finally, what I took to be

Trout's early view of winged dryfly. Note apparent detachment of wings and hook bend mirrored on undersurface.

Troutfly on edge of 'window'. Note 'flare' effect on wings.

Dry fly well supported with hackle. Note stones reflected onto undersurface.

Dry fly lying low in surface film.
Note 'double hook' image.

Trout's view of an emerger type fly. Note 'sparkle' effect.

Trout's view of Starling
Hackle.

number three. This time, after another few casts I hooked the lower trout, much larger than the first, and after a pull and tussle returned him. Not a very appreciative trout, he shot upstream and sent the other two packing. We anglers are too often tempted to move on to greener pastures but patience can often bring quicker and easier rewards.

Knowing where trout are likely to be found is half the battle. Previous experience of a particular river should provide a diary of events. A trout hooked and lost here, one seen but spooked there, a small dimple under the willows, small incidents to be stored for another day. Three essentials are required for a trout's survival. He must have security from predators, a feeding zone consistent with his size and most important a fair abundance of trout food. Wherever trout are found in such prime feeding spots, they move only if forced by larger trout, floods, or the urge to spawn. They are not unduly adventurous creatures. Should a good trout be located, but alarmed one day, you may be assured he will not be far away on your next visit. A trout's domain is generally no more than 20 or so metres upstream or downstream of his station. This should give a considerable advantage in locating the fish on a later visit.

Lake trout, have similar territorial rules but are much more mobile and to stray too far beyond his 'beat' may mean a confrontation with his neighbour. These areas may seem ill defined but a lake trout will identify some lake bed objects as boundaries. The angler if patient and unobserved, can plan an ambush. However, it doesn't always work out. Some time ago, my son Norm and I were driving along the terraces above Lake Wakatipu when he spotted a fish cruising near the shore. It was action stations and

Obvious.

Master of disguise.

after scrambling down through matagouri scrub, we reached the lake edge and, following the fish, waited for him to turn. Eventually we gave up and returned to the car. We carried on for another three kilometres and once again waited for him. Sure enough in a short while he appeared, still travelling at roughly ten knots. We never did catch up. What his rush was and where he was going was, and still is, a mystery.

If fishing is a bit slack on the main river and a backwater is nearby then that is where to find a trout. Waylaying cruising trout is an exciting business, especially in quiet ponds where the angler may have a grandstand view. Such waters are wonderful places to observe the habits of the trout and especially his repertoire of hunting skills — swerving suddenly to grasp an unfortunate bully or rising nymph, or brushing the weed for boatmen or damselfly nymphs. The still waters provide him with many surface-trapped insects while the mud contains countless numbers of bloodworms or midge larvae. It is fascinating to watch a large trout ploughing a mud bed to disturb such insects and then select victims from the ensuing cloud. If I had my choice of trout situations it would be sitting quietly, in cover by a backwater and watching a series of dimples as a fish slowly approached from under his sanctum of drooping willows. In those cases, instead of pursuing the trout, the roles are reversed. After casting a small dry or unweighted nymph close to his 'beat' it becomes a waiting game. There is no rod waving or water disturbance to alert him but it does need nerves of steel as he rises to the floater or swerves to take a twitched nymph.

If you want to increase your chances of catching a trout then find him first. This is, of course, very time consuming but it is time well spent. Trout camouflage is one of nature's success stories. Whatever their surroundings they seem to be able to merge with them. It has something to do with eyesight, rods and cones, and melanin and carrotin. Whatever, they are certainly masters of disguise. Recently I stumbled across a thin, near-black trout, of some three pounds. He was an example of what occurs when trout are near life's end, when eyesight fails. With it goes the ability to blend with surroundings. If they are caught it is an act of mercy to despatch them. In this case I decided to cast a nymph in front of the fish. He ignored it, as he did with many other offerings, but when I decided to end the game and bent to catch him he took off with a burst of speed a two-year-old rainbow would have been proud of. So much for failing eyesight!

With moderate skills almost anyone can observe trout in calm, not too deep, water. It is when the surface is ruffled by wind or current that trout spotting becomes not only difficult but sometimes intuitive. Stalking trout can become almost an art form. To watch an experienced stalker do his stuff is fascinating. What impresses the most is his fluidity. On some river sections he will move with almost indecent haste, in others, almost painfully slowly. That is because he eliminates parts of the river where trout are unlikely to be found — known as 'thin' water. He concentrates on water levels between knee and thigh depth, quiet areas just downstream of willows, bank edges under willows, that defined strip between fast and slow currents, behind, and sometimes in front of large boulders or logs, against prominent river bed stones that trout often use as boundary markers. And a particular favourite lie of trout — the 'eye' of the pool. Another prime area is wherever an upper pool dumps into the lower. This is the equivalent of a 'lunch counter' to the trout and as such should be thoroughly fished before moving on.

A new chum will always look for the familiar shape of a trout. So does the experienced fly fisher but, in addition, on the slightest suspicion, he will look for movement. It may be nothing more than a momentary disruption of the stream bed pattern, a vague flicker of a black-tipped tail. The movement of an upstream trout will suggest another is near. One of the most difficult places to locate trout is in mottled shadow. They blend in so perfectly that even talented 'spotters' have great difficulty. Again the only way to discover the fish is to watch for movement.

On long clear flats locating trout should prove no problem. The

immediate area should be thoroughly scanned, then, well upstream where the acute angle of vision, gives the river bed a bland appearance. Against that background the darker longitudinal shapes of fish are more easily seen. The main problem in trying to spot trout on flats is glare. While Polaroid spectacles are helpful they do not eliminate glare. In those situations the only way to spot trout is to move away from the stream edge and, by looking back towards trees or a high bank, try to create a dark reflection zone.

Even a metre or so above river or lake level can make all the difference when searching for trout. Providing the angler can remain undetected, high ground is always an advantage. Great care is needed in the approach but once a trout is located the next step is to note his position and match it with a recognisable marker. For instance, a larger than usual boulder or stone, sunken log, coloured stone or an overhanging branch. From a height, a trout may be obvious but at river level the fish will most likely be invisible to the angler. Bankside markers opposite the trout should be noted with his approximate distance from them. Learning to spot trout should be a priority. I have long held a rough rule of thumb. If undetected, an angler can spot a trout, he has a 70% chance of catching him; If not, his chances are reduced to 30%. That little tit-bit emphasises my opening gambit — time spent watching trout is money in the bank.

Typical 'drop off'.

STRATEGY

I know fly fishers who can throw a great line but are only fair trout catchers and others whose casting would sometimes make you weep but they know, to a fine art, how to approach a trout. One such friend would take all afternoon to fish a stretch no longer than a football pitch, searching every square metre of water — standing, crouching, or sitting; waiting for a trout sign. Even nearby he is difficult to spot, from a distance, so well does his khaki/green outfit blend in with the surroundings.

After you've studied a trout's movements, strategy plays a vital role. The fish may be in midstream and well down, in which case, an angled side approach is recommended. With a cast well upstream and from the side the fish should see the nymph before the leader nylon. Because light rays will be refracted the trout will be nearer to you than his image and care should be taken not to line him. Mid-stream trout tend to capture your full attention and it is easy to ignore prime water in the foreground. Inspect it carefully. The times I have nearly stepped on trout while searching upstream!

Not infrequently a trout will be found in the most inaccessible situation, maybe well under tree branches or a handspan above a surface log. The strategy then is to weigh up the time you will waste trying for him and your possible reward. Unless the fish is twice the normal size for the water I find it pays to move on. Apart from the likelihood of a hang-up and a

Where trout are easily 'spooked'.

deflated ego, there is nothing wrong with bypassing a 'too difficult' trout. But be sure to give the same place a good inspection on the way back. He may well have moved to a nearby more vulnerable position.

That happened to me not so long ago when I fished a small South Island stream on a hot summer's day. Running through both pasture and patches of bush it was a pleasant river to fish but it provided plenty of challenges. I arrived at an unusual pool. Normally if you have one high bank you have some beach on the other but in this case I had high banks on both sides and the pool itself was narrow and deep. Lying against the far wall and between clinging clumps of bushes were two trout — a rainbow over 2kg and a metre or so upstream, a brown of around the same weight. On my side I was hemmed in by bankside scrub and branches. There was absolutely no chance of casting from the bank and even less of floating a fly over either fish. But I had one big advantage. Time.

Sooner or later I had a feeling they would move. So I slipped over the bank edge into the nearly chest-deep water. It took quite some time for the rainbow to shift but eventually he settled downstream in the tail of the pool, circled, then came up again. When he came up to my boots I could almost imagine his eyeballs bulging and some fishy expletive. He confirmed that with a blistering dash upstream, fortunately, on my side. The brown continued to rise, little tantalising dimples, just enough to keep me waiting. After what seemed like hours the big fish moved slowly out from the edge of his protecting twigs. The next titbit floating downstream contained a hook which, after a classic rise, needed the slightest of twitches to drive home, and after thrashing about the pool for a while he slipped

Typical feedline.

into the net. I weighed that fish and he went just under 2½kg, a pretty brown who, no doubt, on his return, had lots to say to his pink-striped neighbour.

If there is one thing an angler takes note of when he sets out for his favourite fishing spot it is the wind. For the upstream fly fisher a strong downstream wind can be a problem but it can sometimes be used to advantage. Under a ruffled surface, trout are harder to detect but so is the angler and this makes the fish, once sighted, easier to approach. If the wind is very strong a stiffish rod carrying an extra line weight and a short leader should be used. Casting from the side will prove easier, and, because wind velocity at the surface is less, the rod should be held low. Under those conditions and at river level it will be difficult if not impossible to see the fish and, unless a sighter is used, a well-hackled dry fly gives a better chance. If the end of the leader piles up a little so much the better, it gives more drag-free drift.

The reverse is true for the lake angler who is usually unhappy with bright, cloudless, calm conditions, and who prefers a 'chop', when a fly line and leader are less conspicuous and, if he or she is on a lee shore, trout feeding on food flotsam drifts can provide good sport.

Perfect fishing conditions are rare. Other aspects than wind can affect our fishing and not the least is water temperature which, in turn, often determines the trout's feeding activity. While trout are not harmed by a low temperature they suffer if it gets too high, anything over 30°C (86°F) is considered lethal. Trout are at peak activity at around 17°C (63°F). During high summer and low water conditions they become lethargic through lack of oxygen and if possible will seek tributary outfalls. I saw a good example of that when fishing a small Taupo tributary with John Goddard. It was near the head of a long, bushclad pool that we saw them.

'I've never seen anything like that in my life!' said John. Some 20 large rainbows were lined up against a sheer underwater bluff and only occasionally changing positions. A newcomer would arrive and a resident depart. Throughout our fishing lives we are used to seeing trout facing the current but these were lying *across* the stream with their nebs up to the wall. Discussing it later we came to the conclusion that by some freak of nature a cool current of water was issuing from a fissure in the rock.

In early spring when rivers carry 'snow water', trout tend to stay near the river bed. Although insect life at that time is also at a low ebb it doesn't mean that, given the chance, they will not feed. A weighted nymph trotted well down through the 'eye' of a pool will take fish and for this tactic it is best to use a short leader with a prominent sighter. I have carried a small thermometer for years and often test river temperatures. It takes little effort and gives me a good idea of prospects. Below 10°C (50°F) I know I am going to have to work hard for my fish, probably with a deep nymph or

stonefly imitation and I need to have it, not only on the bottom, but also right on the button. Trout in those cold temperatures tend to save energy and seem reluctant to move out of station. When the mercury hits the 17°C (63°F) mark it is smiles all round and with plenty of insect activity it could well be a dry fly day. If it records high river temperatures I know to pass midday fishing and wait for the evening when the water temperature drops and trout once again become active.

The trout is a complex animal and, although we understand much of his functions and habitat, how *he* observes *his* world will always be a mystery. But for the angler — what a fascinating challenge!

What we all love to see.

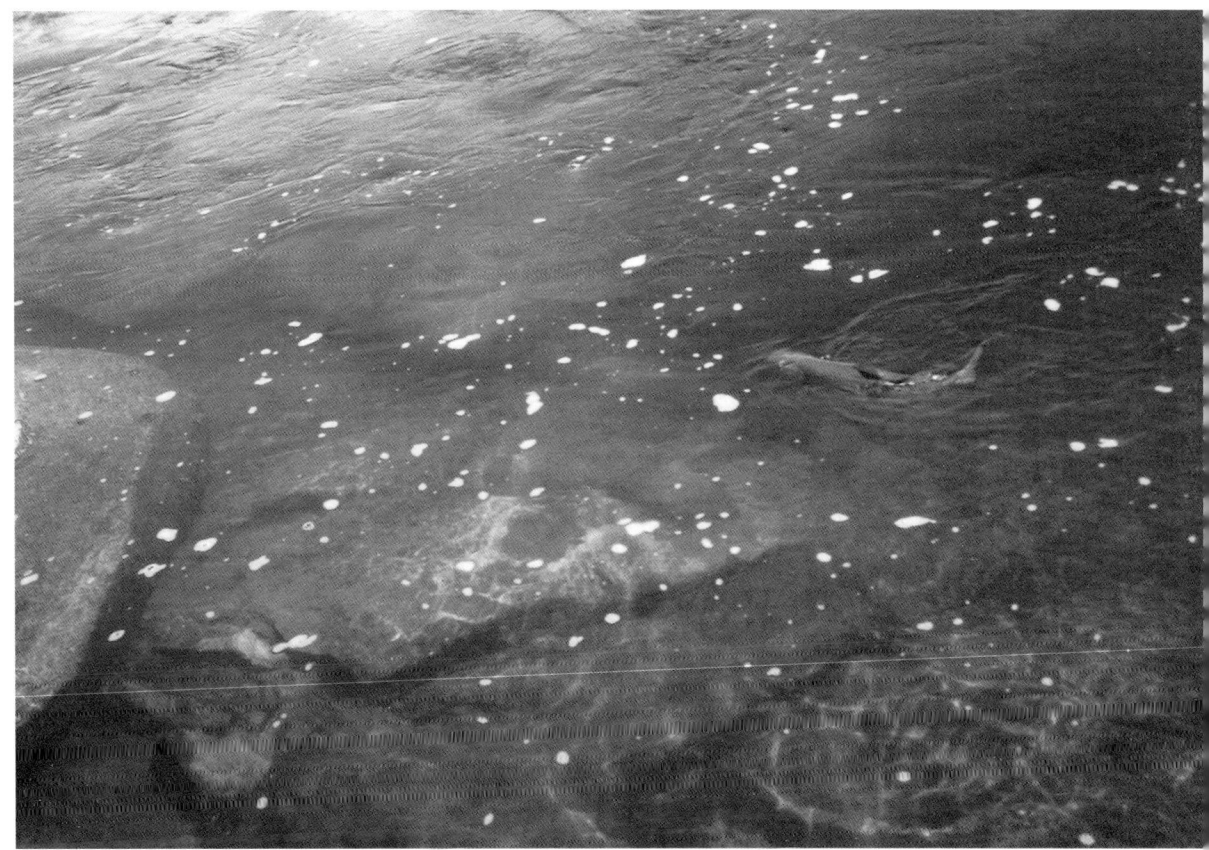

14

Fishing on Top

Dry fly fishing is supposed to be the 'creme de la creme' of angling skills. A long-held tradition is that it needs the skill of a surgeon, the eyesight of a hawk, the patience of Job and the ability to convince one's self that the dry fly man sits on the right side of God. He may do so but it's a very shaky seat! And even if true, on the left hand seat would be a nymph fisherman. Both are equally skilled hunters. The wet fly and lure anglers, skilful and sporting though they may be, are in a different class. They do not normally select targets, but take pot luck. There is no doubt, however, that the ultimate pleasure in fly fishing is to watch a large trout ogle the fly for a second or so then sip it in. The subsequent flurry when he gets the message only adds to the occasion. For a dry fly fisherman the thrill is in the rise.

As in many other endeavours, using proper equipment is half the battle. Using a Taupo style rod with a #8 line on a spring creek may seem ridiculous but I have seen it tried. Recently a visitor arrived with a #5 rod and #8 line which made it a very trying day for both of us.

The first consideration in acquiring a fly rod is to decide where you want to use it. I'm the last person qualified to recommend economy in rods but, even so, one rod will hardly suffice. Two are usually required — one

142

for small streams and medium-sized rivers, another for lakes or big water. The rod size may be dependent on the type of water but the material depends on the size of your pocket. Graphite is today's popular choice but I have seen too many fine anglers with a glass rod in hand to be dogmatic. A graphite rod is light and powerful but glass is not far behind. I own glass fibre rods that perform better than mediocre graphite but in general a graphite will outperform the glass. In low water conditions on my local stream, the Motueka, I often use a 10ft (3m) graphite with a #4 d.t. floater which delivers a very gentle presentation but if the wind gets up I'm in trouble.

Recommended rod sizes

Small hill streams Spring creeks Low country meandering rivers	1-2 Kg trout 24cm/27cm (8/9ft) #5/6 line
Mountain streams Backcountry action rivers Lakes	2-4 Kg trout 27+cm (9ft) #7/8 line

Note: Use will depend to some extent on wind conditions. The 7/8 is better for mastering a strong downstream wind in smaller waters while a 5/6 rod is more pleasant to use on windless days when stalking cruising trout along lake edges.

A delightful stretch on the Motueka.

Ron Harvey lays out a gentle cast.

Conditions

Ideal dry fly fishing conditions are days with a gentle upstream breeze, sunshine, occasional light cloud, normal river flow, and water temperature around 15°C (60°F). For lake edge stalking with a dry fly, a slightly ruffled surface is ideal. Trout have excellent vision, adapted for both water and air. From a metre depth he can see objects on the surface through a two metre circular window, and an angler's vertically held rod from up to ten metres. Even this slim image appearing on the rim of his window is enough to diminish or put paid to your chances. He enjoys binocular vision for a small angle upstream and for the rest a mirror reflection of the stream bed. If you think a trout has missed your approach, think again, and double your precautions.

Locating Trout

On a recent trip with a visitor from England I pointed out a large trout feeding on nymphs in fast, chuckling water. At first glance there were just the flickering shapes of coloured cobble and pebble but before long some of the shapes resolved into the hint of a trout. When the hint slipped sideways we were onto that 70% chance. Only when I was satisfied he could see the fish did I encourage him to offer a Greenwell's. If lacking spotting skills, his casting proved faultless, and the trout took the floater without hesitation. After admiring the big brown he returned it and, much to my delight, before the day was out, was able to locate his own fish.

The Fly v Presentation

Trout may have relatively small brains but they have enough instincts to fool us much of the time. They may be wary but they are usually very hungry, and when trout food insects are within their reach they become rather cavalier. However, providing the angler has done his or her homework, these trout can be readily caught. An often talked about issue among fly fishers is 'which is more important — presentation or the fly?' It goes like this. If you cover a fish well and he refuses your artificial all is not lost and you might try another fly. So when trout are 'on the feed' presentation takes a front seat over choice of fly. But when trout are very selective the choice of fly, both in pattern and size, becomes critical and no amount of excellent casting with the wrong fly will connect you to a trout. In fact the excessive casting is almost sure to put the fish down. To be able to cast competently is not too difficult but to choose the right fly at the right time *and* deliver a good cast is by no means easy. The right combination should be always the goal.

Rise Forms

A rough indication of what trout are feeding on can be gained by studying the rise forms — watery rings that often prove the trout's undoing. Some gurus maintain it is possible to deduce the actual mayfly species by the type of rise form which, in my opinion, is stretching it by a country mile. But a keen observer will know if the rises are produced by trout taking emerging mayflies, mayfly nymphs, or midge pupa. In smooth water cocked mayfly duns are easily seen, even in rough water they can be obvious but except on a mirror surface, detecting spinners can be difficult. In ruffled water it is practically impossible.

Where trout are rising with no surface insects apparent it is a pretty safe bet the trout are taking either battered nymphs or malformed duns in the surface film. In both cases the rise form is a simple dimple. (What a pretty phrase!) A succession of bold rises in broken water often indicates a trout feeding on immature mayfly nymphs. Slashing rises are a fair sign that trout are taking caddis, a rise form more prevalent in the evenings. During high summer dimpling can also be attributed to the presence of willow grub and/or, depending on the area, fruit hoppers.

SURFACE FOOD

Emergers

We have all experienced the frustrations of fishing the long glassy flats when trout are rising to the floating nymphs on the point of emerging as winged adults. Countless numbers of these mature nymphs, or emergers, eager to rid themselves of their underwater cloaks struggle in the surface film but the great majority fail to make it. Perhaps genetically weak, exhausted or battered by currents, the frail insects cannot penetrate the water surface tension and drift slowly down the feedlines. Legions of them provide the trout with a leisurely feast evident by the series of gentle, regular, sips. There is no better fly to imitate these drab nondescript wounded emergers than the soft hackled wet and in particular the Starling Hackle.

Fishing for sippers can be one of the most demanding of fly fishing skills and requires proficient casting with fine tackle. The angle of attack is most important — an approach from behind is by no means the only one. Depending on the river conditions (in many cases it is impossible to reach trout from the rear) an approach from opposite or even upstream of the fish is often successful. Trout feeding on emergers generally move directly up the feedline, turn downstream to a lower position and repeat the performance. Strange behaviour, as they might as well hold station and pick up the nymphs as they pass.

Even if he is satisfied with the imitation there is no disguising the attached nylon. When fishing for sippers the choice of tippet size is of vital importance. It should be as fine a diameter as possible consistent with the size of the trout. In some cases, especially in snag free pools, 6x may be suitable but I am more comfortable with 5 x b.s. On that fine nylon setting the hook should be by tightening only; a savage strike on such fine prestressed nylon is a recipe for disaster. Hooked and played on modern nylon, and if knots are secure, it takes a very big trout to break 5x.

Leader length is most important and under low water conditions should be a minimum of three metres plus a tippet length no less than a metre. Where a fish can be seen, sighters are not only unnecessary but run the risk of alarming the trout. In any case the sip is so obvious you would need to be looking the other way not to see it. With so many naturals in the feedline the chances of your fly being taken are much reduced. But, on the plus side 'sippers' are eager feeders and not easily put off. Fishing for trout taking emergers is full of quiet excitement and not recommended for those with a weak heart.

Midges

How often have we used the expression, 'I'm b------d if I know what they are taking!' This is usually on calm summer evenings after the sun has set. The 'bewitching' hour of poets and dreamers. A pool surface may be literally covered in dimples but the angler's nymph, emerger, or spinner patterns are ignored. One evening, determined to solve the riddle, I did the unthinkable — waded into the rising trout and swept the surface with my gauze net. While quite a few drowned and battered duns and a few spinners were present the bulk of the catch was hundreds of midge pupae. Prior to that I had thought these tiny insects would hardly be worth the trout's attention but, more by luck than judgement, having caught a fish, proved the case with an autopsy. Even so I had little sport during those feeding bouts. My lack of success was depressing and I had given up hope of catching more than the occasional fish on devised patterns. That is until I fished with John Goddard.

On one of those magical evenings, as darkness approached, we fished the head of a large pool, John was silhouetted against the golden sunset with the chuckle of a side stream adding quiet music. That was punctuated by the chatter of his reel as time and time again he played and brought good trout to the net. It wasn't just his creative pattern that provided success but his method of fishing it. Until ready to emerge, midge pupae float vertically in the surface film then, turning to the horizontal, put on a tiny burst of speed before taking off. When fishing for trout very close to the surface the slightest drag is usually fatal but in this case a small movement actually improved the chances. By sometimes raising the rod tip John imparted this characteristic to his suspended midge imitation and by the end of the evening he had not only landed some nice trout but also become a convert to midge fishing.

Williams fishing his deadly emerger.

Spinners

During periods of mayfly emergence, trout dine well, mainly on malformed insects but also on the rising nymphs and the pre-flight duns. The fortunate few that do enter our atmosphere travel quickly to bankside herbage to wait for the final transformation. Depending on the species, this may take from one to two days. Then, in true Cinderella style, they emerge as dainty spinners. Male spinners having done their duty and exhausted from nuptial flights, scatter and die, but the females have a final purpose and return to the river to deposit egg masses. Perhaps because of the Darwinian theory where only the fittest survive and when most feathered predators have roosted, the main body of spinners choose late evening to return to the river. Another possible reason is that, reluctant to lose body fluids during higher daytime temperatures, they postpone the change from dun to spinner until the cooler evening. From working with mayflies in aquariums, and other insect classes, it has been my experience that while aquatic insects enjoy warm conditions they dislike direct heat. That possibility is borne out by the fact that spinner falls occur *throughout* overcast days and not just evenings.

For whatever reason, the return of the female spinners means trout once again have the opportunity for a feast and take station just below the surface. It is important to guess roughly the intervals between rises and present the fly as soon as possible and directly in line with the fish. With many naturals to choose from, trout will rarely stray more than a metre either side of station but if the timing, placing, and spinner choice are correct then an offer is likely. These are very selective trout and it is essential that your imitation, if not slavish, should be similar in size to the natural. This in general would conform #18 hook size but I find that a #16 hook size is more successful, perhaps because it stands out from the herd. A long fine tippet is advised, together with a very lightly dressed spinner pattern.

Random spinners

Although long still pools act as insect collectors (and for that reason we observe spinners much more easily) spinners deposit their cargo of eggs throughout the length of the river and not only in the evening. In fast or broken water, with wings flat on the surface, the delicate mayflies are practically impossible to see. But among the exhausted, dying, or dead spinners riding on the surface, many others are ovidepositing and on their short rides downstream are ready targets for the trout. When spinners are detected diving and dancing over the stream it is time to prospect. Pocket water, those semi-calm areas among rapids, are ideal for that kind of fly fishing where the skill is in placing the artificial spinner to linger in the tiny pools and laybys.

In longer bouncy feedlines and using a short line it is possible to 'bounce' the fly down the faster currents. When held downstream and moved

John Goddard fishing fine
and far off.

upstream in erratic 'flicks' the situation between the trout and the fly sometimes becomes a case of the dog chasing the rabbit. Fly fishers in general abhor a strong downstream wind but here is a case where it is a distinct advantage. By holding the rod erect and letting the leader be blown over the stream and by lowering the rod tip at random, letting the fly skip the surface, it gives a very realistic impression of a dancing spinner. I take great delight in using this method as it gives me the chance to drive a trout wild when usually the reverse is true. If the fly is then allowed to drift a short distance downstream it often results in a very hefty rise. The fly should not be allowed to sink. When spinners are on or over the water the trout's eye is on one place only, the surface.

Willow Grubs

Trout feeding on willow grubs need not necessarily be close to willows. These small yellow larvae also enter the water column and are washed downstream but a waxy coating keeps them on or very near the surface. During the later summer months if trout consistently refuse various artificials it is very likely they are sipping willow grubs. They are very small and very difficult to see in the surface film but where trout are dimpling under the willows it is fairly safe to assume grubs are the flavour of the month. These rises are delicate; the duchess taking tea. A #18 hook dressed with a yellow and green fine wool body with a tiny black hackle and smeared with a touch of Mucilin sometimes does the trick. When feeding on willow grubs these fish are hard to catch — the epitome of selective trout.

Hoppers

When fruit hoppers are in season trout will nearly always be found line astern in the feedline suds. The rises are different from those to the willow grub, gentle but leaving a small swirl as the hoppers are sucked from the surface. These fish are nowhere near as difficult to hook as those feeding on willow grubs. If a good drift is allowed, and very fine nylon used, then trout take imitation hoppers with great confidence. The only problem is that in the ensuing turmoil all trout in the vicinity retire. The pull and tussle of even one good fish is reward enough.

Prospecting

Prospecting with a fly rod is the bread and butter of fly fishing. Trout need food and protection and, although shallow bouldery reaches may harbour many insects, they lack shelter for larger trout. They are not prime trout areas. A good prospector will double the premium fishing time of the beginner by passing by such unfruitful sections of river. No amount of book knowledge can teach this valuable lesson. While the value of spotting trout has been emphasised it should not be at the expense of passing possible trout stations. These may be in fast, turbulent areas and can be searched with random casting through broken stretches.

One of the best proponents of dry fly on broken stretches is my good friend, Vern Williams. His technique is to work the water quickly exploring every likely place with no more than a couple of casts, keeping the fly high on the surface by a raised rod tip and using as short a line as possible. Occasionally he lets the fly dawdle in eddies and doubles his chances by using a nymph as a dropper tied a handspan above the dry. Dry flies, mayfly or caddis, should be of top quality for this type of fly fishing, riding high on the surface and retaining a minimum of moisture.

All the above methods of surface fishing will catch trout depending on river or lake conditions. If trout can be observed feeding then half the battle is won. If there is anything to be learned from experience it is that, with any method, it always pays to fish as fine as circumstance will allow but none more so than when fishing 'on top'.

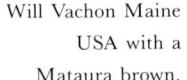

Will Vachon Maine USA with a Mataura brown.

15

In Between

Around the turn of the century dry fly fishing was in its infancy and the most popular methods were wet fly or lure fishing. Although early records show some dry fly fishing in the South Island it was uncommon. Nymph fishing, as we know it today, was practically unheard of but the balance has now swung in the other direction. Today, on your travels, you will rarely find a dedicated wet fly fisherman. It seems the art has, if not disappeared, then is well under the bank. But not quite. I know a few fly fishermen who are skilled in the use of the wet fly, and when conditions are right, use it to good effect.

Unlike the stable English chalk streams which rely on underground aquafiers, our rivers are at the mercy of the elements and can quickly change from shrunken pools to roaring floods. It is during the inbetween periods, when the river levels are falling and the water is tinted, that the downstream wet fly comes into its own. What is a 'wet' fly, and what is it supposed to represent? It can hardly represent a natural insect when fished at relatively high speed down and across the current. Used in that way it can only be an attractor, taking advantage of the trout's natural aggressiveness, or perhaps its strong territorial instinct. With wings tied parallel and close to the body the artificial may well appear to the trout as a small fish.

151

Compared to upstream stalking with dry fly or nymph, wet fly fishing, may seem a 'chuck and chance' affair requiring the minimum of skill. There is more to it than that. To be successful in wet fly fishing the angler needs to picture the underwater world from a trout's viewpoint, to be able to anticipate the action of the current on the fly and, most of all, to be able to 'read' the water. The ability to anticipate where a good trout is lying is the hallmark of a good wet fly fisherman.

Unlike lure fishing, large flies are not necessarily the best. Trout have exceptionally keen eyesight and although a trout fly may seem small to us, its relative size to the fish is much larger. I rarely use wet flies larger than a #12 with a #14 used in most situations. On the other hand I have never found use for anything smaller. Hackled wets are generally the equal of winged versions but in faster water I find the latter most successful, no doubt because of the more distinct silhouette. Whether to use more than one fly on the leader is a matter of choice. A trout may take a single wet fly without hesitation yet find a multi-fly cast unnatural. Current flow should be taken

Graeme Marshall returning a
Karamea heavyweight.

into consideration. Wet flies rarely sink much below the surface and in quiet water the 'wake' caused by a team of flies may well scare fish. But when bouncing down riffles teams of wet flies can be very effective. As the old farmer said when he whacked the donkey between the eyes with a four by two, 'You have to get their attention!'

In fishing downstream with a wet fly wading is almost a necessity. Knee depth wading is usually sufficient with the cast made slightly upstream and across. This allows the fly to drift and sink for up to a metre before the current takes hold. A golden rule for wet fly fishing is that you either fish deep with weighted fly or sunken line or fish the wet close to the surface. Trout, unless in transit, rarely inhabit mid-water sections. Depending on water speed, the line should be 'mended' by throwing a 'belly' upstream. This allows the fly an even longer drift and the wet fly time to sink. When the current takes hold and the fly starts to swing away from the central current the rod should be held as high as practical. The fly is now at its most attractive position and a 'strike' can be expected. Often this is just a quiet 'pull'; the fish just nipping the fly. That is one of the disadvantages of downstream wet fly where, unlike the take of an upstream fish and the hook pulled *into* his jaws, it is now, in effect, an upstream pull and more likely pulled *out* of them. But, providing the hook is kept sharp, and this is most important, enough trout do take hold to make wet fly downstream productive and, when a big brown is on the end of the line in fast water, very satisfying.

One of the nicest aspects of downstream wet fly is its therapeutic value. There are times when hunting trout is more important and others where the languid casting is in tune with your mood. Sidling down the river, five steps and a cast, anticipation, a little manipulation of the fly, the swing,

Mike Steed on the Mangles.

the smooth recovery and the repeat, all fluid, make wet fly fishing a leisurely affair. There is time to appreciate your surroundings and value them. Quite likely, fantails keep you company while further downstream the terns could be taking mayflies on the wing. A swirl against the far bank goes into the memory bank, the old labrador wades up for a cuddle and suddenly there is a strong pull of a good fish. Wet fly fishing at its best.

The other way is to fish wet fly upstream. There is a pool not far from my home that is perfect for it. Willows line the far bank where deeper water brushes the tree roots and cuts out safe shelters for large trout. A long glide flows from the pool above, spreads across a wide pebble apron and over a wide gravel fan. From there the knee-deep water ripples over a long stretch of cobblestones. Fish lie against the willows but by far the best fishing is to wade very slowly up the pool, angle casting upstream with a soft hackled fly.

A few days ago, on my way upstream, I detoured down a little gravel track to the pool. It was mid-morning, the sun was well on its way and a slight breeze augered well for the day's fishing. I wasn't the first angler on the pool. Disturbed from a downstream log a white-throated shag flapped his ungainly way upstream. Chaffinches fluttered back and forth across the river, a sure sign of insect activity but these were not the focus of my attention. Scores of little swirls along the full length of the pool revealed that a mayfly hatch was in progress. I hurriedly set up the rod while eyeing a few duns. These were ignored so I knew the swirls were made by trout taking semi-hatched duns, the intermediate stage between nymph and adult mayfly. Trying on a Partridge and Orange I began casting upstream and across, concentrating on some particular fish. Stroking the hackled wet through the swirls brought results. Soon came that lovely thrill of a trout hooked.

And so it went on — cast, drift, tighten, play, and release. Each fish brought to hand; the fly, worse for wear, twisted out and the fish returned. If one can break off fishing during a major hatch (great willpower required) to examine the current, many tiny brown shells or shucks will be seen. These are the discarded cases of small brown mayfly nymphs. Others may still be attached to the actual nymph in the last stages of its underwater existence, struggling to be rid of its clinging coffin. They are easy prey for trout. Duns that break free of the surface film and float for a short distance before taking flight, are usually ignored. This is not surprising as trout rarely waste energy; the nymphs are much easier to secure.

A wet fly is just as effective when fishing small streams. Using a short line the angler is in constant contact with the fly, and a sighter is unnecessary. It is a searching process where every possible holding spot for trout is probed with a single, soft hackled fly. It is an active process, drawing the fly through quiet eddies where helpless insects have gathered, or dancing it through riffles or behind large boulders where trout, sheltering from pressure of current, can take advantage of passing food. During summer and late autumn, when caddis abound, a variation of the hackled wet can make for an exciting evening.

It was during a warm December evening on the Eglinton that I found myself covered with adult caddis. It was the biggest match of those clumsy insects I have ever encountered and I was frequently brushing them from my face. Up to that time (around 7pm) there had been little activity apart from an occasional rise but then the splash of a large trout broke the stillness of the evening. More rises followed. I quickly changed to a sedge pupa imitation, which is really an adapted soft-hackled wet, and cast to the nearest fish rising against the far bank. At least a score of fish were splashing on the surface but I could not get one to respond. Just before I resorted to a change of fly the penny dropped. Caddis pupae have a habit of spurting through the surface film before taking off and, as soon as I realised this and gently pulled the fly to the surface, I was in business. Another half hour of daylight would have seen a few more hefty Fiordland fish hooked but as darkness closed I returned to camp, with the tantalising splash and thump of active trout ringing in my ears. Events like that are magical. As you walk back down along the river's edge, you look upon the snow-tipped mountains touched with a last rosy glow and, as the light from the campervan appears, you really wonder if you have already parted from this mortal coil, and are indeed, in a fly fisher's Paradise.

Going home.

16

Down Below

Only in recent years have fly fishers recognised that surface insects play but a minor part in the trout's diet. Previously trout flies were either fished as floaters or as 'wet flies', cast down and across the stream or trailed behind a boat. But times have changed and artificial nymphs now take up much space in the fly box. This change is in no small way due to Tony Orman's *Trout with Nymph* (1974, Hodder & Stoughton). On reading that excellent book many fly fishers became more aware of the relatively untouched larder of the river bed.

Overseas, one of the first real proponents of nymph fishing was G.E.M. Skues a one-eyed (physically) solicitor who practised in Victorian London and who fished the chalk streams of southern England, particularly the Itchen. On Hampshire and Surrey chalk streams, only dry fly fishing was allowed, and 'wet' fly fishing on those hallowed waters brought down the wrath of members and ensured banishment from those exclusive clubs. Skues was no exception. His advocacy of nymph fishing made him a pariah among many of his fellow members. He ended his days fishing in relative obscurity. As a tribute to this quiet but determined character I would like to quote from *My Sporting Life* by J.W. Hills:

. . . that emancipated young woman, the nymph . . . nymphs donned their scanty skirts at an earlier age, but Skues, with a truly modern mind, taught them how to make up their faces, and showed them that the less clothes they put on, the more attractive they would be. That, I think, is the nymph in a nutshell.

Unlike the English chalk streams, with their regular and sometimes prolific hatches of mayfly and caddis, our rivers and streams are for the most part, and for want of a better term, 'wilder'. They are not, apart from some spring creeks, gentle-flowing, crystal-clear streams, strewn with pretty water weeds, a haven for aquatic insects. More of the rough and tumble variety where trout feed for the most part on subsurface insects. And not only are our rivers wilder but so are the trout. Most English streams, lakes and reservoirs are stocked with trout from fish farms, mostly rainbows, and they cannot be compared with our own wild fish.

My first excursion into nymph fishing was back in the early sixties while fishing the Clinton River in Fiordland, a fairyland of large browns and rainbows where a trophy fish could well be around the corner. Apart from sporadic tramping parties on the Milford Track one saw few people and could fish the bush-clad turquoise pools in absolute privacy. But one day I got a surprise. Rounding the bend I saw an angler with rod hooped, well attached to a very large fish which, despite some searing runs and leaps, was finally brought to the net. By this time I was close enough to witness the return of a rosy-flanked rainbow but not before I saw him twist the large nymph from its jaws. That incident not only enlightened me as to the effectiveness of nymph fishing but seduced me into the study of aquatic insects.

If the mayfly is the 'prima donna' of the trout stream then the nymph provides the full orchestra. Trout could well exist without adult mayflies but they would find it difficult to survive without a regular diet of nymphs. And not only mayfly nymphs. The main source of a trout's diet is subsurface which makes fishing the underwater nymph, caddis or other trout food imitation by far the most productive. There is also much to be gained from studying the habits of trout and particularly nymphing fish. The majority of food items either drift near the river bed or at the surface and because these food items are relatively small, it is almost impossible to ascertain whether they are mayfly nymphs, caddis larvae or various other items. Fortunately, trout feeding actively on or near the stream bed are not too choosy and will usually take most artificial nymphs if well presented.

A rough rule to determine which imitation to use is related to the water characteristics. Large, slow-flowing rivers and quiet streams tend to favour the use of smaller nymphs whereas the more boisterous backcountry rivers host much larger naturals. But this is only a general rule. Backcountry trout, owing to easier access and increased pressure from guided visitors, are

becoming just as fussy as their downcountry counterparts. Trout in the lower stretches of rivers similar to the Mataura and the Motueka will often shy at nymphs tied on hooks larger than #12s and, during low-water summer conditions, even #14s. Getting to know which trout stream insects inhabit a river or lake can save much time in selecting an appropriate nymph. For example on the West Coast spring creeks, where *Colorburiscus* is common, the Kakahi Queen Nymph is a good choice. In sluggish limestone rivers, it would be a poor one.

A good idea when nymph fishing strange water is to quarter the river, casting upstream over the near half then across to the opposite side and flipping the line belly upstream to give maximum dead drift. No more than two casts into one spot is a good maxim for nymph fishers. Trout have an excellent view of underwater insects or artificials and on a drag-free drift, and if your nymph is not taken on the first two casts, you should carry on quartering. Obviously there are some sections of rivers that carry more fish than others and time spent finding these is money in the bank. Favourite trout holding spots are 'drop offs' where the upstream pool dumps into the lower creating that peculiar gravel bench.

This is almost like a serving dish to the trout which, with little effort, can intercept the insect goodies. In pool 'eyes', where friction currents form a quiet area, trout can conserve energy yet be ready and alert for passing food. These are the most productive areas for the nympher as are pockets of relatively calm water behind boulders or logs. Other favourite trout lies are slab rock seams which afford shelter from strong currents but are handy to food. In the deeper water along a high bank trout feel secure while taking

Nymph fishing on the
Otapiri, Southland.

food from an underwater feedline. Fishing this type of water, dead drift, is much easier if you are fishing across from shallows, a situation that allows good line control. Nymphing techniques vary depending on the type of water you fish. The dead drift method is the most favoured, but sometimes other methods pay off.

I recall only a few weeks ago when fishing the upper Rai, a small Marlborough stream. It was a lovely day, sunny but not too hot. The birds were singing, there was a slight breeze upstream, and the little river sparkled as it danced its way through willow lined pools. After parking the Honda and trudging downstream for maybe two kilometres I started back, casting here and there while searching for a familiar shape. Despite my optimism I had little success and it was within sight of the bike that I had my first chance. In a small pool I spotted a flickering shape in the riffle. A large brown lay on the edge and from his movements he was obviously nymphing. After perhaps a half hour (at least!) pitching the nymph above him I was no nearer hooking that fish. Convinced that there was no fault with the nymph I moved cautiously upstream until opposite the fish and tried again. On the first cast he lifted and took the nymph. Previously, in casting directly upstream, in fast, shallow water, the nymph had little drift time before being swept downstream. This, to the trout, was obviously an unnatural speed. The side cast allowed only a short drift but at a pace natural enough to seduce him. He weighed over two kilograms and paid for his mistake by posing for a photo before going back home.

Nymph fishing is an art in itself. It needs the same casting skills as in fishing the dry but requires more understanding of the habits of aquatic insects. As many a fly fisher will testify, when taking surface insects trout can be extremely selective but when feeding subsurface they have more catholic tastes. The trick is in knowing when to go subsurface. Successful nymph fishing depends on getting the nymph down quickly, giving the trout every opportunity to inspect it. The best way to accomplish this is by the dead drift method, casting the nymph well upstream of the fish, giving maximum time for it to sink. An upstream cast should be to either side of the trout which gives him the opportunity to see the nymph before the nylon. False casting (kept to a minimum) should be well away. The nymph should reach the trout naturally and should not be pulled. Conversely, when fishing in deeper water for unseen fish, it is an advantage to lift the nymph *at the end of the drift* to simulate a rising nymph or caddis pupa.

Fishing the nymph can be productive throughout the season but especially so after a fresh or minor flood. Providing snow water is not present and temperatures are around 10-12 °C (50-54 °F) trout will still feed randomly on nymphs, caddis and stonefly larvae. During high water, nymphs and caddis larvae are in a constant battle with strong currents and a shifting stream bed. Even during moderate flows many lose their purchase and

The Angler's Companion.

provide the trout with a regular supply of food.

A floating line and a long fine leader are the best combination to obtain drag-free drifts, the line riding high on the surface while the business end of the leader cuts quickly through the water. Another option is the 'sink tip' line which seems ideal but in fact anchors the leader and defeats the purpose by quickly dragging the nymph. A leader of between three and four metres is very practical, the tippet size depending on water conditions with 3x (2.0kg) not too heavy for spring conditions.

Casting a weighted nymph is nowhere near the delightful sensation as when using a dry fly but by throwing a more open loop it is not unpleasant. Another word for casting a nymph is 'pitching', when the nymph is delivered much slower; almost a 'lob' and more in the style of a roll cast. False casting is not normally necessary.

While it is essential to apply some weight to the nymph or leader, too heavy a nymph can invite tangles. One small #8 split shot placed approximately 15cm from the nymph is usually sufficient to sink the nymph or small, paper-thin lead strips can be cut from beaten lead and pinched off as required. A recent innovation is 'lead' paste, hard when cold and soft with finger heat. My first encounter with this 'stuff' was when nymph fishing with Gary Borger on the Motueka. I found the material excellent. A pinch wrapped around a leader knot was easy to apply and easy to remove.

The river was high, not flooded but full enough to make trout spotting very difficult. In the fast water between pools, even from high banks, practically impossible. Only in the pools did I stand a chance but even then I'd tramped a long way before I spotted not one but three good browns, a pretty sight, all at the head of the long narrow pool and feeding. It didn't take long to scare the lower fish and the second one followed soon after. The reason being that they were very deep in fast current and my repeated efforts failed because neither were prepared to lift to the nymph which I'd increased in size to a #10, my green stonefly pattern. The third and largest fish seemed destined to join the other two when I saw the solution. Reduce the size of nymph but weight it with a small pinch of lead putty. This time the nymph was deep enough to tempt him and when I saw him lift for the first time I set the hook. He should have weighed over five pounds but, still thin from over wintering, he would be a pound short. Unless used in very fast water, almost torrent grade, very big nymphs do not act naturally and hard fished trout are not easily deceived. I'm a great believer in using as small a nymph as practical and this is where a tiny piece of lead putty pinched around the hook eye often makes all the difference. As I removed the hook prior to releasing him I noticed the tear mark of another hook, indicating that, probably the day before, (it was the second day of the season) another angler had released him in this catch and release section of the Eglinton.

Whereas the dry fly fisher knows, or should know to a nicety, when to tighten on a duped fish, the nymph fisher's warning signals are much more subtle. To hook a fish rising conventionally to a dry fly needs only a pause before striking or an even longer pause if the fish is facing downstream. The position of the fly and rise can be seen and the timing judged with near accuracy. In slow-flowing water of reasonable depth the taking of a nymph may also be easy to see, especially the 'flash' of the trout's jaws, or if unseen, the 'twitch' of the tippet. In this case it is not too difficult to hook the fish. But in more rapid water and at varied depth, it is necessary to use a sighting aid. It is of little use if trout takes the nymph and then is given time to eject it. In heavy or fast water no tell tale 'twitch' of the leader can be relied on. Strike indicators are essential.

These can range from a small piece of cork or pith from a quill feather, or just a twist of wool tied above a leader knot. Most nymphers prefer fluorescent wool or acrylic fibre and enhance its buoyancy with fly floatant. Too big an indicator can easily alert shallow lying trout and should be trimmed accordingly. The distance from nymph to sighter is not critical but should be judged according to the depth of water fished, ranging from tippet knot to the leader butt section. With experience the right size indicator can be matched to the weight of the nymph and, by bouncing the artificial just clear of the bottom, an angler stands a good chance of success. The occasional snag assures you the depth is just right. Fishing upstream to observe fish can be most rewarding but in the right circumstances fishing a 'random' nymph can produce just as much excitement. A recent angling aid is 'float paste', a fluorescent yellow, almost unsinkable dough that also hardens on contact with cold water and is easily removed.

Deeper lying trout have a wider all-round vision and just because the trout is nymphing doesn't allow for a less cautious approach. The diameter of the tippet still needs to be as fine as possible. The cast to a nymphing fish, to reach his depth, may have to be one or even two metres above him — the nylon drifting downstream can easily contact the fish. Or give him a clear view of the tippet. With a trout near the surface, ripples may help hide the nylon and the nymph may be pitched just upstream of him. As in most aspects of fly fishing there is an element of luck but, providing the nylon is fine enough, trout can be duped. In deeper, rougher water, nylon size is not as critical as one may expect perhaps larger fish, and 3X or 2X tippet points may not be out of the way. I have yet to see trout waters where it is essential to use larger than 2X. With today's light but powerful and flexible rods and the quality of present-day nylon the odds are in the angler's favour.

Mayfly nymphs are not the trouts only underwater targets. On hot summer days, when even a keen angler, seeks shade there is good fishing to be found along lake margins and backwaters. A walk along the edge

Paul Farrow with magnificent rainbow.

(creeping is much better) will reveal myriads of panicking boatmen or, if weedy, damselfly nymphs swimming between green fronds. The trout are well aware of that larder and often cruise across those shallows in search of a titbit or two. Creatures of habit, trout usually reappear on schedule. They are also extremely shy, a fact that makes casting over them a waste of time. It requires an ambush. It is not difficult to figure a backwater trout's routine and after choosing a well-camouflaged position, cast a #16 Pheasant Tail or Waterboatman to intercept him. Fine nylon and absolute stillness are required until the fish is about an arm's length from the nymph, when the rod tip is gently raised. Backwater trout are usually big and it is the highlight of any fly fisher's experience to see, at close quarters, the formidable jaws open and the nymph sucked in.

Lake edge fishing can be just as exciting and can produce some surprises. It was raining heavily as we drove down the West Coast bush but by the time we reached Lake Ianthe the clouds parted and the lake seemed so inviting we decided to stop there for the night. After dinner we 'dimped up' and prepared for a short walk along the lake edge. 'Taking the rod?' asked Jean. 'I don't think so,' I replied, then recalled other occasions when I'd had good cause to regret those words. With the departing rain clouds now glowing with orange and fiery reds we enjoyed a spectacular sunset. Late arriving waterfowl etched the mirrored surface with black and gold ripples. But it was a series of approaching rings that caught my attention. Within easy casting distance, they came ever nearer and with fumbling fingers I tied on a damselfly imitation and cast it in front of the approaching fish. It was too dark to see the fish when a great swirl appeared on the surface I pulled up, tight. Then I wondered what I had struck.

'It's a monster!', exclaimed Jean after I had been playing the unseen fish for near ten minutes. Jean timed it. Far out into the deeps he ran, the backing quickly following the fly line, and I was left with no choice but to hang tight to the last coils. After what seemed an hour I finally brought the fish within reach. In the last rays of the sun we saw the fish slide into the net and I lifted it triumphantly. So are our hopes often dashed. No trophy this fish, but a deep-flanked six-pounder, with the fly firmly attached to his dorsal fin! When you think of the odds against the fly hooking the fish again after being dislodged from his jaws the mind boggles. I must have looked a little sad. 'Well', said Jean, 'at least you can sleep tonight without wondering whether you lost the "fish of a lifetime"'. I must admit, I did sleep that night. But not very well.

In river fishing an upstream approach is almost essential but sometimes a bit of lateral thinking is required. That situation occurred when I was fishing a North Island river with the local wildlife officer. It was picture postcard water with toi toi framing the blue/green pools. A Maori village would have completed the idyllic setting but the only Maori in the vicinity was a very large fellow up to his chest in a downstream pool. He was trailing a cord attached to which were some very large rainbows.

A lonely ostracized brown lay deep in the next pool and, after some pestering, disappeared into nearby weeds. Further upstream we had to detour around scrub and arrived back on the river opposite another deep pool. Two browns were at the head and both well down. They completely ignored my upstream casts but looking downstream were three or four large rainbows hugging the bottom. Backcasting into the head of the pool I allowed the nymph to sink well down then let it swing up from the bottom. Before the rest of the fish got the message I had hooked and landed two rainbows very near the 2kg mark. No fancy casting, no crafty stalking. A classic case of choosy browns and cavalier rainbows showing that if there is one certain factor in trout fishing it is the uncertainty.

A Last Cast

With over half a century of fly fishing behind me I have seen many changes in methods, attitudes and the quality of trout fishing. Increasing pressure on trout waters has made for more discerning fish, especially as a season progresses. As world populations grow and more affluent societies appear and with the inevitable increase in tourism, great care will be needed to protect our relatively clean, fresh, trout streams from overfishing. Modern techniques and more effective fishing tackle also take a toll. On the plus side a new 'catch and release' generation augers well for the future.

There is still much to be learnt in the proper technique of releasing trout with undue harm. One of my greatest riverside delights is to see a young fly fisher gently hold his trout in the net, wish him well, then let him go. Fishery managers are now more enlightened, and, except in rare circumstances, are sceptical on re-stocking, placing more emphasis in enhanced habitat. Pollution now presents our greatest threat not the least that from badly planned forestry with hillside scars allowing silt and fine gravels run-off to choke mainstream river beds. Mining and the drowning of trout streams for small hydro schemes follow a close second.

But, providing we keep constant vigil against the commercialisation of our sport and are quick to act if threatened, there is every hope that our young anglers will continue to enjoy our wonderful trout fishing and protect it.

More power to their rods.

Norman Marsh
'Streamside'
1995

Index